Error Correction Codes for Non-Volatile Memories

Error Correction Codes
for Non-Volatile Memories

R. Micheloni, A. Marelli and R. Ravasio

Qimonda Italy srl, Design Center Vimercate, Italy

 Springer

Rino Micheloni
Qimonda Italy srl
Design Center Vimercate
Via Torri Bianche, 9
20059 Vimercate (MI)
Italy
rino.micheloni@qimonda.com

Alessia Marelli
Qimonda Italy srl
Design Center Vimercate
Via Torri Bianche, 9
20059 Vimercate (MI)
Italy

Roberto Ravasio
Qimonda Italy srl
Design Center Vimercate
Via Torri Bianche, 9
20059 Vimercate (MI)
Italy

ISBN: 978-1-4020-8390-7   e-ISBN: 978-1-4020-8391-4

Library of Congress Control Number: 2008927177

Printed on acid-free paper

9 8 7 6 5 4 3 2 1

springer.com

Contents

Preface

In the 19-th century there was a revolution in the way people communicate: the telegraph. Although the user would see it as a device "transmitting words", it did not, not in the sense letters do, which has been the only established way to communicate for centuries. The telegraph sends two signals: a short beat and a long beat. The way these beats correspond to letters and hence to words is the famous Morse code.

Other revolutionary ways to communicate soon arrived, like the telephone and the radio, which could use the same principle: adapt our language to new transmission media via a suitable transformation into signals. An important part of this transformation process is the use of a code and a common feature is the presence of disturbance affecting the quality of transmission.

Several ad-hoc methods were studied for different kinds of media, but it was the stunning contribution by Shannon[1] which initiated a rigorous study of the encoding process in order to reduce significantly the disturbance (and to use the media efficiently).

After his paper, thousands of papers and dozens of books have appeared in the last 60 years. This new subject, "Coding Theory" as it is commonly called today, lies in the fascinating interface between Engineering, Computer Science and Mathematics. Unsurprisingly, contributions by mathematicians have focused on the design of new codes and the study of their properties, while contributions by engineers have focused on efficient implementations of encoding/decoding schemes. However, most significant publications have appeared on IEEE Trans. on Inf. Th., where a clever editorial policy has strongly encouraged fruitful collaborations between experts in different areas.

The astounding amount of research in Coding Theory has produced a vast range of different codes (with usually several alternative encoding/decoding algorithms), and their availability has naturally suggested new applications. Nowadays it is hard to find an electronic device which does not use codes. We listen to music via heavily encoded audio CD's, we watch movies via encoded DVD's, we travel on trains which use encoded railway communications (e.g., devices telling the trains when to stop and when to go), we use computer network where even the most basic communication enjoys multi-stratified encoding/decoding, we make calls with our mobile phones which would be impossible without the underlying codes, and so on.

---

[1] C. E. Shannon, "A Mathematical Theory of Communication" in *Bell System Tech. J.*, Vol. 27, pp. 379-423 623—656, 1948.

On the other hand, the growing application area poses new problems and challenges to researchers, as for example the need for extremely low-power communications in sensor networks, which in turn develop new codes or adapt old ones with innovative ideas.

I see it as a perfect circle: applications push new research, research results push new applications, and vice versa.

There is at least one area where the use of encoding/decoding is not so developed, yet. Microchips and smart cards are the core of any recent technology advance and the accuracy of their internal operations is assuming increasing importance, especially now that the size of these devices is getting smaller and smaller. A single wrong bit in a complex operation can give rise to devastating effects on the output. We cannot afford these errors and we cannot afford to store our values in an inaccurate way.

This book is about *this* side of the story: which codes have been used in this context and which codes *could* be used in this context. The authors form a research group (now in Qimonda), which is, in my opinion, the typical example of a fruitful collaboration between mathematicians and engineers. In this beautiful book they expose the basic of coding theory needed to understand the application to memories, as well as the relevant internal design.

This is useful to both the mathematician and the engineer, provided they are eager for new applications. The reading of the book does suggest the possibility of further improvements and I believe its publication fits exactly in the *perfect circle* I mentioned before.

I feel honored by the authors' invitation to write this preface and I am sure the reader will appreciate this book as I have done.

Massimiliano Sala
Professor in Coding Theory
and Cryptography Dept. of Math.
University of Trento

Acknowledgements

After completing a project like a technical book, it is very hard to acknowledge all the people who have contributed directly or indirectly with their work and dedication.

First of all, we wish to thank Elena Ravasio, who has taken care of the translation of the text from Italian to English.

We thank Andrea Chimenton, Massimo Atti and Piero Olivo for their special contribution (Chap. 5) on the reliability aspects of the floating gate Flash memories.

We also wish to thank Prof. Massimiliano Sala who wrote the foreword for this book.

We have to thank Mark de Jongh for giving us the possibility of publishing this work and for his continuous support. Special thanks to Adalberto Micheloni for drafts review.

Last but not least, we keep in mind all our present and past colleagues for their suggestions and fruitful discussions.

Rino, Alessia and Roberto

1 Basic coding theory

1.1 Introduction

In 1948 Claude Shannon's article "A Mathematical Theory of Communication" gave birth to the two twin disciplines: information theory and coding theory. The article specifies the meaning of efficient and reliable information and, there, the very well known term "bit" has been used for the first time. Anyway, it was only with Richard Hamming in 1950 that a constructive generating method and the basic parameters of Error Correction Codes (ECC) were defined.

Hamming made his discovery at the Bell Telephone's laboratories during a study on communication on long telephone lines corrupted by lightening and crosstalk. The discovery environment shows how the interest in error-correcting codes has taken shape, since the beginning, outside a purely mathematical field.

The success of the coding theory is based precisely on the motivation arisen from the manifold practical applications. Space communication would not have been possible without the use of error-correcting codes and the digital revolution has made a great use of them. Modems, CDs, DVDs, MP3 players and USB keys need an ECC which enables the reading of information in a reliable way.

The codes discovered by Hamming are able to correct only one error, they are simple and widely used in several applications where the probability of error is small and the correction of a single error is considered sufficient.

In 1954 Reed and Muller, in independent ways, discovered the codes that today bear their names. These are able to correct an arbitrary number of errors and since the beginning they have been used in applications. For example, all the American Mariner-classes deep space probes flying between 1969 and 1977 used this type of codes.

The most important cyclic codes, BCH and Reed-Solomon, were discovered between 1958 and 1960. The first ones were described by Bose and Chaudhuri and through an independent study by Hocquengheim; the second ones were defined by Reed and Solomon a few years later, between 1959 and 1960. They were immediately used in space missions, and today they are still used in compact discs.

Afterwards, they stopped being of interest for space missions and were replaced by convolutional codes, introduced for the first time by Elias in 1955. Convolutional codes can also be combined with cyclic codes. The study of optimum convolutional codes and the best decoding algorithms continued until 1993 when turbo codes were presented for the first time; with their use communications are much more reliable. In fact, it is in the sector of telecommunications where they have received greater success.

A singular history is that of LDPC (Low Density Parity Checks) codes first dis-covered in 1960 by Gallager, but whose applications are being studied only today.

This shows how the history of error-correcting codes is in continuous evolution and more and more applications are discovering them only today. This book is de-voted to the use of ECC in non volatile memories.

In this chapter we briefly introduce some basic concepts to face the discussion on correction codes and their use in memories.

1.2 Error detection and correction codes

The first distinction that divides the family of codes is between the detection ca-pability and the correction capability.

A s-errors detection code is a code that, having read or received a message cor-rupted by s errors, is able to recognize the message as erroneous. For this reason a detection failure occurs when the code takes a corrupted message for a correct one.

A very easy example of an error detection code able to recognize one error is the parity code. Having a binary code, one extra bit is added to the message as an exclusive-OR among all the message bits. If one error occurs, this extra bit won't be the exclusive-OR among all the received bits anymore. In this way a single er-ror is detected. If two errors occur, the extra bit isn't able to recognize errors and the detection fails. For this reason the parity code is a single error detection code.

On the other side an error correction code is a code that is not only able to rec-ognize whether a message is corrupted but it also identifies and corrects the wrong positions. Also in this case there is a correction failure, in other words the code is unable to identify the wrong positions.

Suppose that we want to program one bit. We shall build a code in the follow-ing way:

- if we want to program a 0 we write 000;
- if we want to program a 1 we write 111.

When reading data, the situations depicted in Table 1.1 can occur.

Table 1.1. Read and decoded data for the error correction code described above

Number of errors	Read data	Decoded data
0 errors	000	0
	111	1
1 error	00$\underline{1}$	0
	0$\underline{1}$0	0
	$\underline{1}$00	0
	11$\underline{0}$	1
	1$\underline{0}$1	1
	$\underline{0}$11	1

If the read data are 000 or 111 no error occurs and the decoding is correct. The other combinations show the case of one error in the underlined position. With the use of a majority-like decoding we can see that the code is able to correct all the possible one-error cases in the right way. Now suppose that 2 errors occur. Suppose that we programmed 0, encoded as 000, but we read 101. In this case the majority-like decoding mistakenly gives the result 111, so that the programmed bit is recognized as 1.

Summarizing we can say that the code is corrector of one error, detector of 2 errors but its disclosing capacity, that is the detection capability without performing erroneous corrections, is of 1 error.

The use of error correction instead of error detection codes depends on the type of application. Generally speaking, it is possible to say that:

- We use a detection code if we presume that the original data is correct and can therefore be re-read. For example if the error in a packet of data occurs on the transmission line, with a detection code we can simply ask its re-transmission.
- If the frequency of such event increases over a certain limit, it may be more efficient in terms of band to use a correction code.
- If the error is in the source of the transmitted information, e.g. in the memory, we can only use a correction code.

Hereafter we shall deal with error-correcting codes, that is, those able to detect and correct errors.

1.3 Probability of errors in a transmission channel

Error correction theory is a branch of information theory that deals with communication systems.

Information theory treats the information as a physical quantity that can be measured, stored and taken from place to place. A fundamental concept of information theory is that information is measured by the amount of uncertainty resolved. The uncertainty, in mathematical science, is described with the use of probability.

In order to appreciate the benefic effect of an error correction code on information, it is necessary to briefly speak about all the actors and their iterations in a communication channel. A digital communication system has functionalities to execute physical actions on information. As depicted in Fig. 1.1, the communication process is composed by different steps whereas the error correction code acts in only one.

First of all, there is a source that represents the data that must be transmitted. In a digital communication system, data is often the result of an analog-to-digital conversion; therefore, in this book, data is a string made by 0s and 1s.

Then there is a source encoder which compresses the data removing redundant bits. In fact, there is often a superfluous number of bits to store the source information. For compressed binary data, 0 and 1 occur with equal probability. The source encoder uses special codes to compress data that will not be treated in this book.

Sometimes the encrypter is not present in a communication system. The goal is hide or scramble bits in such a way that undesired listeners would not be able to understand the information meaning. Also at this step, there are special algorithms which deal with cryptography theory.

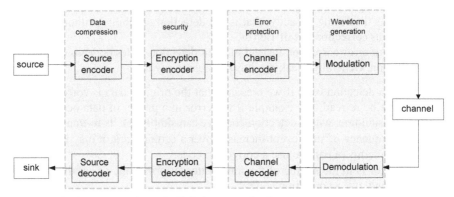

Fig. 1.1. Block representation of a digital communication system

The channel encoder is the first step of the error correction. The channel encoder adds redundant bits to the information, so that, during decoding process, it is possible to recognize errors that occur in the channel. The main subject of this book is the study of the different methods used by the channel encoder to add those bits and by the channel decoder to recognize error positions.

The modulator converts the sequence of symbols from the channel encoder into symbols suitable for the channel transmission. A lot of channels need the signal transmission as continuous-time voltage or an electromagnetic waveform in a specified frequency band. The modulator delivers its suitable representation to the channel. Together with the modulator block it is possible to find codes that accomplish some particular constraints as maximum number of all 1 string allowed and so on.

The channel is the physical means through which the transmission occurs. Some examples of channel are the phone lines, internet cables, optical fibers lines, radio waves, channels for cellular transmissions and so on. These are channels where the information is moved from one place to another. The information can also be transmitted in different times, for example writing the information on a computer disk and reading it in a different time. Hard disks, CD-ROMs, DVDs and Flash memories are examples of these channels and are the channels this book deals with.

Signals passing through the channel can be corrupted. The corruption can occur in different ways: noise can be added to the signal or delays can happen or signals can be smoothed, multiplied or reflected by objects during the way, thus resulting in a constructive and/or destructive interference patterns. In the following, binary symmetric channel and white Gaussian noise will be considered. Channels can take different amount of information that can be measured in capacity C, defined as the information quantity that can be carried in a reliable way. It is in this field that it is possible to find the famous Shannon's theorem, known as channel coding theorem. Briefly the theorem says that, supposing the transmission rate R is inferior to the capacity C, there exists a code whose error probability is infinitely small.

Then, the receiving signal process begins. The first actor is the demodulator, which receives the signal from the channel and converts it into a symbol sequence. Generally, a lot of functions are involved in this phase such as filtering, demodulation, sampling, frame synchronization. After that, a decision must be taken on transmitted symbols.

The channel decoder uses redundancy bits added by the channel encoder to correct errors that occurred.

The decrypter removes any encryption.

The source decoder provides an uncompressed data representation.

The sink is the final data receiver.

As already said, errors occur in the channel. As a consequence, the understanding of the error probability of a binary channel with a white Gaussian noise, helps us to understand the effect of an error correction code.

Thus, suppose we have a binary data sequence to be transmitted. One of the most used schemes for modulation is the PAM technique (Pulse Amplitude Modulation), which consists in varying the amplitude of transmitted pulses according to the data to be sent. Therefore the PAM modulator transforms the binary string b_k at the input into a new sequences of pulses a_k, whose values, in antipodal representation are

$$a_k = \begin{cases} +1 & \text{if } b_k = 1 \\ -1 & \text{if } b_k = 0 \end{cases} \tag{1.1}$$

The sequence of pulses a_k is sent as input of a transmission filter, producing as output the signal $x(t)$ to be transmitted (Fig. 1.2).

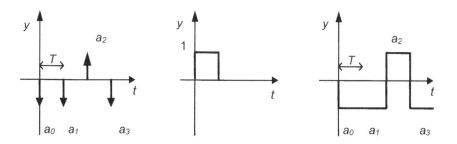

Fig. 1.2. Representation of transmitted pulses, single pulse waveform and resulting waveform

The channel introduces an additive, white and Gaussian noise $w(t)$ which is summed to the signal $x(t)$ in order to have the received signal $y(t)$

$$y(t) = x(t) + w(t) \qquad (1.2)$$

The received signal is sampled and the sequence of samples is used to recognize the original binary data b'_k in a hard decision device that uses the threshold strategy as represented in Fig. 1.3.

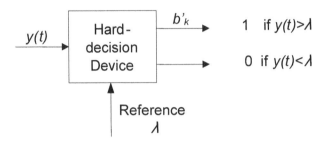

Fig. 1.3. Hard decision device with threshold reference

The highest quality of a transmission system is reached when the reconstructed binary data b'_k are as similar as possible to the original data b_k.

The effect of the noise $w(t)$ introduced by the channel is to add errors on bit b'_k reconstructed by the hard-decision device. The quality of the digital transmission system is measured as the average error probability on bit (P_b) and is called Bit Error Rate or BER. The error occurs if:

$$\left(b_k = 0 \text{ and } b'_k = 1\right) \text{ or } \left(b_k = 1 \text{ and } b'_k = 0\right) \qquad (1.3)$$

The best decision, as regards decoding, is taken with the so-called maximum a posteriori probability criteria: we decide $b=0$ if the probability of having transmitted "0" by having received y is greater than the probability of having transmitted "1" by having received y:

$$P(b = 0 \mid y) > P(b = 1 \mid y) \qquad (1.4)$$

In case of transmission of symbols with the same probability it is equivalent to the maximum likelihood decoding, i.e. we decide $b=0$ if the probability of having received y by having transmitted a "0" is higher than the probability of having received y by having transmitted a 1. That is, Eq. (1.4) becomes

$$P(y \mid b = 0) > P(y \mid b = 1) \qquad (1.5)$$

By hypothesis, the noise $w(t)$ introduced by the channel is Gaussian with mean 0, so the probability density functions f, mathematically described in Eq. (1.6), are two Gaussians with means $-m$ and $+m$ and variance σ^2 as shown in Fig. 1.4.

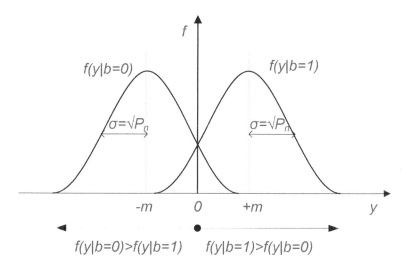

Fig. 1.4. Representation of received signals afflicted by noise

$$f(y \mid b = 1) = \frac{1}{\sqrt{2\pi P_n}} e^{\frac{(y-m)^2}{2P_n}}$$

$$f(y \mid b = 0) = \frac{1}{\sqrt{2\pi P_n}} e^{\frac{(y+m)^2}{2P_n}}$$

(1.6)

The decision criteria, according to Fig. 1.4, is based on a reference threshold $\lambda=0$ and is the maximum likelihood criteria:

- we choose $b'=1$ if $f(y|b=1)>f(y|b=0)$ i.e. if $y>0$;
- we choose $b'=0$ if $f(y|b=1)<f(y|b=0)$ i.e. if $y<0$.

Errors occur in the dashed area shown in Fig. 1.5.

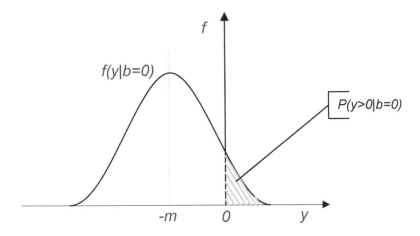

Fig. 1.5. Error probability for a hard decision device

We have:

$$
\begin{aligned}
P_b &= P(b'=1, b=0) + P(b'=0, b=1) = \\
&= P(b=0)P(b'=1 \mid b=0) + P(b=1)P(b'=0 \mid b=1) = \\
&= \frac{1}{2} P(b'=1 \mid b=0) + \frac{1}{2} P(b'=0 \mid b=1) = \\
&= P(b'=1 \mid b=0) = \\
&= P(y > 0 \mid b=0) = \\
&= \int_0^\infty f(y \mid b=0) \cdot dy = Q\left(\frac{m}{\sigma}\right)
\end{aligned}
\tag{1.7}
$$

Therefore, the Q function, evaluated in the ratio between half the distance of two possible signal levels and the standard deviation, is the probability of error for a binary transmission with a hard decision device based on a threshold reference. In the following section we will see how an error correction code is able to deal with this probability.

1.4 ECC effect on error probability

This book deals with error correction codes applied to Flash memories. As already said, in the transmission channel represented by memories, the information isn't physically shifted from one place to another, but it is written and read in different times.

Different reasons for errors (see Chap. 5) can damage the written information so that it could happen that the read message is not equal to the original anymore.

The reliability that a memory can offer is this error probability. This probability could not be the one that the user wishes. Through ECC it is possible to fill the discrepancy between the desired error probability and the error probability offered by the memory. The error probability the decoder uses is the one related to the transmission channels described in the previous section and can be written as

$$p = \frac{Number\ of\ bit\ errors}{Total\ number\ of\ bits} \tag{1.8}$$

It is necessary to clarify the meaning of desired error probability: in order to understand this, it is useful to explain the architecture of a memory (Fig. 1.6).

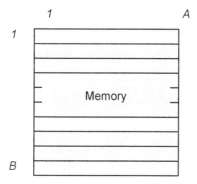

Fig. 1.6. Logical representation of a memory

The memory chip is formed by B separate blocks. Every block is made up by A bits. The desired error probability can be ascribed to the whole memory chip or to the number of error bits regardless of their topology in the chip. Two different parameters are commonly used. The Chip Error Probability (*CEP*) is defined as:

$$CEP(p) = \frac{Number\ of\ chip\ errors\,(p)}{Total\ number\ of\ chips} \tag{1.9}$$

while the Bit Error Rate (*BER*) is defined as:

$$BER(p) = \frac{Number\ of\ bit\ errors\,(p)}{Total\ number\ of\ bits} \tag{1.10}$$

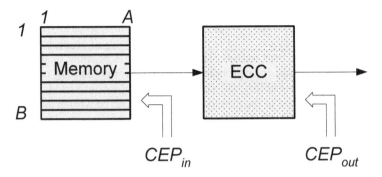

Fig. 1.7. Representation of a system with block ECC. The memory is logically divided into B blocks of A bits. The CEP observation points are highlighted

In a system with a block correction code (Fig. 1.7) the CEP_{in} is given by:

$$CEP_{in}(p) = 1 - (1-p)^{AB} \qquad (1.11)$$

where A is the number of bits per block and B is the number of blocks in a chip. Assuming to use a correction code of t errors, the CEP_{out} is

$$CEP_{out}(p) = 1 - (1-b)^{B} \qquad (1.12)$$

where b is the failure probability of the block to which the ECC is applied

$$b = 1 - [P_0 + P_1 + P_2 + ... + P_t] \qquad (1.13)$$

P_i is the probability of i errors in a block of A bits

$$P_i = \binom{A}{i} p^i (1-p)^{A-i} \qquad (1.14)$$

In conclusion we have that

$$CEP_{out}(p) = 1 - \left[(1-p)^A + \binom{A}{1} p(1-p)^{A-1} + ... + \binom{A}{t} p^t (1-p)^{A-t} \right]^B \qquad (1.15)$$

Figure 1.8 shows the graphs of CEP_{in} (indicated in the legend as "no ECC") and CEP_{out} as a function of p for a system with 512 Mbit memory, 512 Byte block and ECC able to correct 1, 2, 3 or 4 errors.

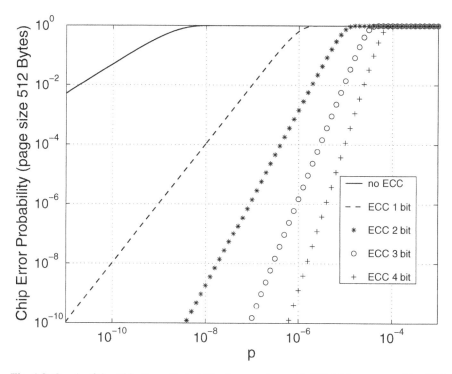

Fig. 1.8. Graph of the Chip Error Probability for a system with 512 Mbit memory. The *CEP* parameters for correction codes of 1, 2, 3 and 4 errors are given

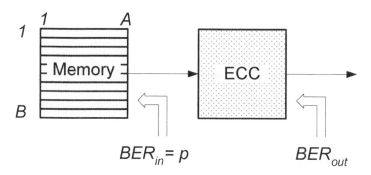

Fig. 1.9. Representation of a system with block ECC. The memory is logically divided into *B* blocks of *A* bits

In a system with a block correction code (see Sect. 1.5) the BER_{in} coincides with p (Fig. 1.9). Assuming the use of a correction code of t errors, the BER_{out} is given by:

$$BER_{out}(p) = \frac{Number\ of\ residual\ erraneous\ bits(p)}{Total\ number\ of\ bits} \tag{1.16}$$

The number of residual erroneous bits, after the correction performed by the ECC, is a quantity experimentally measurable and detectable by simulation. The exact analytical expression of such quantity depends on the capability of the code to recognize an erroneous message without performing any correction and it is generally too complex. It is possible to estimate BER_{out} in defect (BER_{inf}) and in excess (BER_{sup}) using some approximations.

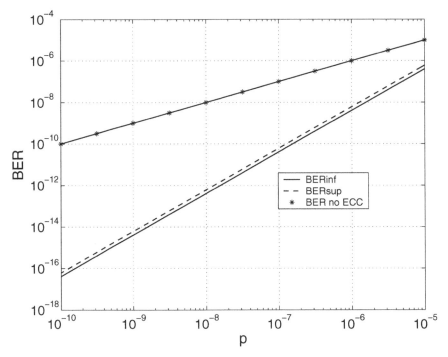

Fig. 1.10. Graph of the Bit Error Rate for a system with ECC corrector of one error and correction block (A) of 128 bits

The calculation of BER_{inf} hypothesizes that the ECC used is able to correct t errors and to detect all the blocks with more than t errors without performing any correction. With this hypothesis, the code never introduces further errors in the block.

$$BER_{out} \geq BER_{inf} = \frac{1}{A}\sum_{i=t+1}^{A} iP_i \tag{1.17}$$

where P_i is the probability that the block contains i errors (Eq. (1.14)).

The calculation of BER_{sup} assumes that the ECC used is able to correct t errors and, when the number of errors is greater than t, it is able to detect it but, trying some corrections, it erroneously introduces other t errors.

$$BER_{out} \leq BER_{sup} = \frac{1}{A} \sum_{i=t+1}^{A} (i+t) P_i \qquad (1.18)$$

In conclusion:

$$\frac{1}{A} \sum_{i=t+1}^{A} i P_i \leq BER_{out} \leq \frac{1}{A} \sum_{i=t+1}^{A} (i+t) P_i \qquad (1.19)$$

Figure 1.10 shows the superior and inferior limits for a code corrector of one error in a block of 128 bits. Figure 1.11 gives the graphs of BER_{sup} for codes correctors of 1, 2, 3, 4 or 5 errors in a block of 4096 bits.

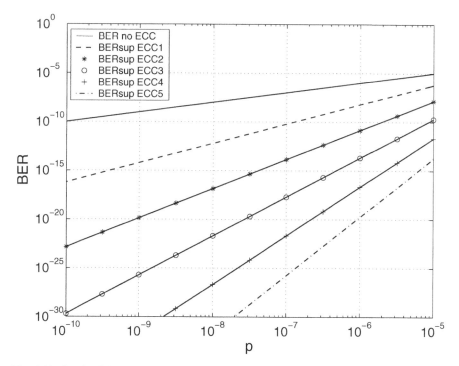

Fig. 1.11. Graph of the approximation of the Bit Error Rate (BER_{sup}) for a system with ECC corrector of 1, 2, 3, 4 or 5 errors and correction block (A) of 4096 bits

1.5 Basic definitions

The object of the theory of error correction codes is the addition of redundant terms to the message, such that, on reception, it is possible to detect the errors and to recover the message that has most probably been transmitted.

Correction codes are divided into two well separated groups: block codes and convolutional codes. The first ones are so defined because they treat the information as separated blocks of fixed length, in contrast with convolutional codes which work on a continuous flow of information.

A second distinction is between the error sources. In channels without memory, noise influences every symbol transmitted in an independent way. An example is the binary symmetric channel that receives exactly two symbols: 0 and 1. The channel has the property that with probability q a transmitted bit is received correctly and with probability $p=1-q$ it is received incorrectly. In this case transmission errors occur randomly in the received sequence. Channels without memory are called channels with random errors. Optical channels or semiconductor memories are two examples.

In a channel with memory, noise is not independent from transmission to transmission. There are two states, a "good" state where transmission errors occur with a probability that is almost 0 and a "bad" state where transmission errors are very likely. The channel is in the good state for most of the time, but sometimes it changes to the bad state. As a result, transmission errors appear in packets called burst. Channels with memory are called channels with burst type errors. Some examples are radio channels, magnetic recordings or compact discs.

1.5.1 Block codes

It has been said that block coding deals with messages of fixed length. Schematically (Fig. 1.12), a block m of k symbols is encoded in a block c of n symbols $(n > k)$ and written in a memory. Inside the memory, different sources may generate errors e (see Chap. 5), so that the block message r is read. The block r is then decoded in d by using the maximum likelihood decoding strategy, so that d is the message that has most probably been written.

A *Code C* (Fig. 1.13) is the set of codewords obtained by associating the q^k messages of length k of the space A to q^k words of length n of the space B in an univocal way.

A code is defined as *linear* if, given two codewords, also their sum is a codeword. When a code is linear, encoding and decoding can be described with matrix operations.

Definition 1.5.1 G is called *generator matrix of a code C* when all the codewords are obtainable as a combination of the rows of G.

Each code has more than one generator matrix, that is all its linear combinations.

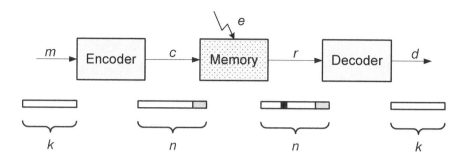

Fig. 1.12. Representation of coding and decoding operations for block codes

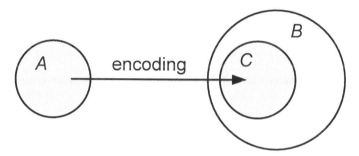

Fig. 1.13. Representation of the space generated by a code

Each code has infinite equivalent codes, i.e. all those obtained by permutation or linear combination of the matrix G.

Definition 1.5.2 Given a generator matrix G of a code $C[n,k]$, each set of k columns of G that are linearly independent is called *information set of C*.

Each information set built by each matrix G' of C is the same as the one built by G.

A code can also be described by parity equations. Suppose we have a binary code $C[5,3]$ whose generator matrix G is:

$$G = \begin{pmatrix} 1 & 0 & 0 & 1 & 1 \\ 0 & 1 & 0 & 0 & 1 \\ 0 & 0 & 1 & 1 & 1 \end{pmatrix} \tag{1.20}$$

Considering the first three positions as an information set, we can define the redundancy positions as functions of them.

Let $a=(a_1, a_2, a_3, a_4, a_5)$ be each vector in C and suppose we know the information positions a_1, a_2, a_3. The redundancy positions can be calculated as follows:

$$a_4 = a_1 + a_3 \tag{1.21}$$

$$a_5 = a_1 + a_2 + a_3 \tag{1.22}$$

Each word in C satisfies these equations, which can therefore be used to write all the vectors of C.

Definition 1.5.3 A set of equations that gives redundancy positions in terms of information positions is called *parity equations set*.

It is possible to express all these equations as a matrix. In the example:

$$H = \begin{pmatrix} 1 & 0 & 1 & 1 & 0 \\ 1 & 1 & 1 & 0 & 1 \end{pmatrix} \tag{1.23}$$

The matrix H is called *parity matrix* for a block code.

Therefore, with reference to Fig. 1.12, encoding a data message m consists in multiplying the message m by the code generator matrix G, according to Eq. (1.24).

$$c = m \cdot G \tag{1.24}$$

Definition 1.5.4 G is called in *standard form* or in *systematic form* if $G = (I_k, P)$, where I_k is the identity matrix $k \times k$ and P is a matrix $k \times (n - k)$. If G is in standard form then the first k symbols of a word are called information symbols.

In the preceding example G is in standard form and the first 3 bits are information bits.

Theorem 1.5.1 If a code $C[n,k]$ has a matrix $G = (I_k, P)$ in standard form, then a parity matrix of C is $H = (-P^T, I_{n-k})$ where P^T is the transpose of P and it is a matrix $(n - k) \times k$ and I_{n-k} is the identity matrix $(n - k) \times (n - k)$.

Systematic codes have the advantage that the data message is visible in the codeword and can be read before decoding. For codes in non-systematic form the message is no more recognizable in the encoded sequence and it is necessary to have the inverse encoding function to recognize the data sequence.

Besides, for each non-systematic code it is possible to find an equivalent systematic one, therefore, as regards block codes, those systematic result to be the best choice.

Definition 1.5.5 The code rate is defined as the ratio between the number of information bits and the codeword length. Given a linear code $[n,k]$ the ratio k/n is defined as *code efficiency*.

Table 1.2 shows the theoretical efficiency for binary codes correctors of 1, 2 and 3 errors in relation to the length of the code n and to the correction capability t.

Table 1.2. Relationship between n, k, t and the theoretical efficiency for codes corrector of 1, 2 or 3 errors

n	k	$n-k$	t	k/n
3	1	2	1	0.33
7	4	3	1	0.57
7	2	5	2	0.29
15	11	4	1	0.73
15	8	7	2	0.53
15	5	10	3	0.33
31	26	5	1	0.84
31	22	9	2	0.71
31	18	13	3	0.58
63	57	6	1	0.90
63	52	11	2	0.82
63	47	16	3	0.75
127	120	7	1	0.95
127	114	13	2	0.90
127	108	19	3	0.85
255	247	8	1	0.97
255	240	15	2	0.94
255	233	22	3	0.91
511	502	9	1	0.98
511	494	17	2	0.97
511	486	25	3	0.95

Definition 1.5.6 If C is a linear code with parity matrix H, then $x \cdot H^T$ is called *syndrome* of x.

All the codewords are characterized by a syndrome equal to 0.

The syndrome is the main actor of the decoding. Having received or read the message r (Fig. 1.12), first of all it is necessary to understand if it is corrupted by calculating:

$$s = x \cdot H^T \tag{1.25}$$

There are two possibilities:

- $s=0 \Rightarrow$ the message r is recognized as correct;
- $s \neq 0 \Rightarrow$ the received message contains some errors.

In this second case we use the maximum likelihood decoding procedure, that is we list all the codewords and we calculate the distance between r and the codewords. Therefore the vector c that has most likely been sent is the one which has the smallest distance from r. At this point it is fundamental to define the distance and therefore the concept of metric.

Definition 1.5.7 It is called *minimum distance* or *Hamming distance d* of a code, the minimum number of different symbols between any two codewords.

We can see that for a linear code the minimum distance is equivalent to the minimum distance between all the codewords and the codeword 0.

Definition 1.5.8 A code has *detection capability v* if it is able to recognize all the messages, containing *v* errors at the most, as corrupted.

The detection capability is related to the minimum distance as described in Eq. (1.26).

$$v = d - 1 \tag{1.26}$$

Definition 1.5.9 A code has *correction capability t* if it is able to correct each combination of a number of errors equal to *t* at the most. The correction capability is calculated from the minimum distance *d* by the relation:

$$t = \left[\frac{d-1}{2} \right] \tag{1.27}$$

where the square brackets mean the floor function.

Definition 1.5.10 A code has *disclosing capability s* if it is able to recognize a message corrupted by *s* errors without trying a correction on them.

The disclosing capability is related to the distance by the following relation:

- if *d* is odd the code does not have disclosing capability;
- if *d* is even $s = t + 1$.

Sometimes the disclosing capability can be much better than in these relations, as described in Chap. 10. These relations regard the minimum disclosing capability and are equal to the maximum disclosing capability in the case of perfect codes.

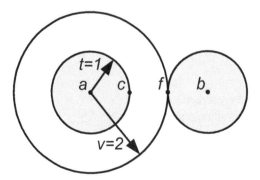

Fig. 1.14. Representation of the space generated by a code with minimum distance equal to 3

Figure 1.14 shows the space of a code with $d = 3$. In this case a and b represent two codewords, whereas c and f represent words containing 1 or 2 errors respectively. In this example the code is able to correct one error: since the word c is in the "correction circle" $(t = 1)$ of the word a it will therefore be corrected with the nearest codeword.

The code is also able to detect two errors: in fact, the word f is in the "detection circle" $(v = 2)$ and it does not represent any codeword. Since all the methods of correction are based on the principle of the maximum likelihood, the code will try to correct f with b, thus making an error. For this reason we can say that the code has disclosing capability $s=1$.

Figure 1.15 shows the space of a code with $d = 4$. In this case a and b represent two codewords, while c, f and g represent words containing 1, 2 or 3 errors respectively. The correction capability t of the code is the same as before; consequently, if the word c containing one error is read, it is corrected with a. When two errors occur and the word f is read, the code recognizes the errors but it does not perform any correction: the disclosing capability of the code s is equal to 2. In case 3 errors occur, therefore reading g, the code detects the errors but carries out erroneous corrections.

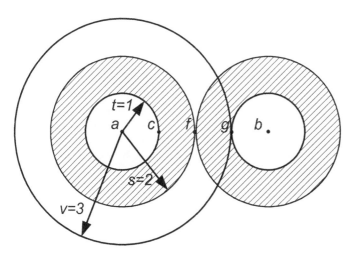

Fig. 1.15. Representation of the space generated by a code with minimum distance equal to 4

Between the length n of the codewords, the number k of information symbols and the number t of correctable errors, there is a relationship known as Hamming inequality:

$$\sum_{i=0}^{t} \binom{n}{i} \cdot (q-1)^i \leq q^{n-k} \tag{1.28}$$

Substantially, the number of parity symbols *(n − k)* has to be sufficient to represent all the possible errors. A code is defined as *perfect* if the Eq. (1.28) is an equality. In this case all the vectors of the space are contained in circles of radius

$$t = \left\lceil \frac{d-1}{2} \right\rceil \qquad (1.29)$$

around the codewords. Perfect codes have the highest information rate of all the codes with the same length and the same correction capability.

From Definition 1.5.10 it is clear that, if possible, a code with an even minimum distance is preferable than a code with an odd one, even if this does not influence its correction capability, since it allows a greater disclosing capability. The possible operation to increase the minimum distance is the extension.

Definition 1.5.11 A code *C[n,k]* is *extended* to a code *C'[n+1,k]* by adding one more parity symbol.

Generally, for binary codes, the additional parity bit is the total parity of the message. This is calculated as *sum modulo 2* (XOR) of all the bits of the message, accordingly it results to be:

- 0 if the number of 1s in the message is even;
- 1 if the number of 1s in the message is odd.

We can observe that, by itself, this type of code, called parity code, is enough to detect, but not to correct, an odd number of errors.

Looking at Fig. 1.15, it is possible to see that, when a double error message is read, it is not possible to confuse two errors with a single error because the extra parity bit remains zero detecting an even number of errors.

In a lot of applications there are external factors not subject to error check which determine the length permitted to an error correction code. Non volatile memories, for example, operate on codewords that have a length power of 2.

When the "natural" length of the code is not suitable it is possible to change it with the shortening operation.

Definition 1.5.12 A *C[n,k]* is *shortened* into a code *C'[n–j,k–j]* by erasing *j* columns of the parity matrix.

Observe that both the shortening operation and the extension are applicable only to linear codes.

1.5.2 Convolutional codes

Convolutional codes have been widely used in applications like satellite communications. Their success is probably due to the simplicity of the decoding algorithm for maximum likelihood, easily hardware parallelizable because of its recursive structure.

The greatest difference in comparison with block codes is that the encoder has a memory and this means that, at every clock pulse, the bits encoded depend not only on the k–th information bit but also on the preceding m ones. Besides, a convolutional encoder converts the whole information string, without taking into account the length, into a codeword. In this way not all the codewords have the same length.

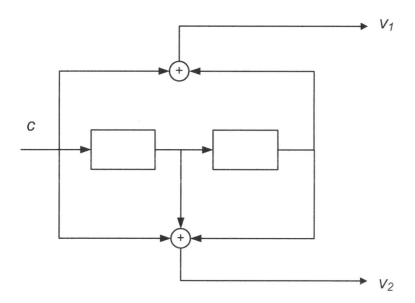

Fig. 1.16. Example of convolutional code (2,1,2)

In Fig. 1.16 an example of a convolutional code is represented. The squares are memory elements. The system is ruled by an external clock that produces a signal every t_0 seconds, the effect is that the signals move in the direction of the arrows toward the following element. The output therefore depends on the current input and on the two preceding ones.

Given an input $c = (..., c_{-1}, c_0, c_1, ..., c_l, ...)$ where l is a second of time, the outputs are two sequences v^1 and v^2: $v^1 = (..., v^1_{-1}, v^1_0, v^1_1, ..., v^1_l, ...)$, $v^2 = (..., v^2_{-1}, v^2_0, v^2_1, ..., v^2_l, ...)$. At time l the input is c_l and the output is $v^l = (v^1_l, v^2_l)$ with:

$$v^1_l = c_l + c_{l-2} \tag{1.30}$$

$$v^2_l = c_l + c_{l-1} + c_{l-2} \tag{1.31}$$

where the sum is meant as binary addition.

Consequently the code is linear and with memory 2. The codeword is formed as:

$$v = \left(\cdots, v_{-1}^1, v_{-1}^2, v_0^1, v_0^2, v_1^1, v_1^2, \ldots, v_l^1, v_l^2, \cdots\right) \tag{1.32}$$

Definition 1.5.13 We define as *impulse answers* or *sequence generators*, the sequences g_j^i generated as *i*–th output of the encoder at inputs $c = (1,0,0,\ldots)$.

Therefore, in the example, we have: $g^1 = (101)$ and $g^2 = (111)$ which are the coefficients of the outputs v^1 and v^2. The coding equations can be written as:

$$v^1 = c * g^1 \tag{1.33}$$

$$v^2 = c * g^2 \tag{1.34}$$

where the symbol "*" is the discrete convolution, that is:

$$v_l^j = \sum_{i=0}^m u_{l-i} g_i^j \tag{1.35}$$

Suppose, for example, that the input is $c = (1011100\ldots)$. The two output sequences will be:

$$v^1 = c * g^1 = (1011100 \cdots) * (101) = (1001011 \cdots) \tag{1.36}$$

$$v^2 = c * g^2 = (1011100 \cdots) * (111) = (1100101 \cdots) \tag{1.37}$$

The transmitted sequence will be:

$$v = (11,01,00,10,01,10,11, \cdots) \tag{1.38}$$

The sequences of generators g^1 and g^2 are the basic elements of the matrix G described in Eq. (1.39).

$$G = \begin{pmatrix} g_0^1 & g_0^2 & g_1^1 & g_1^2 & \cdots & g_m^1 & g_m^2 & \cdots \\ 0 & 0 & g_0^1 & g_0^2 & \cdots & g_{m-1}^1 & g_{m-1}^2 & g_m^1 & g_m^2 \\ \vdots & & & & & & & \vdots \\ \cdots & & & & & & & \cdots \end{pmatrix} \tag{1.39}$$

As for the block codes, the encoding is calculated as $v = c \cdot G$. Called D the delay operator, v and c can be written as:

$$c(D) = \cdots + c_{-1} D^{-1} + c_0 + c_1 D + c_2 D^2 + \cdots \tag{1.40}$$

$$v^i(D) = \cdots + v^i_{-1}D^{-1} + v^i_0 + v^i_1 D + v^i_2 D^2 + \cdots \qquad (1.41)$$

Sequence generators, also known as *generator polynomials*, can be written as polynomials using the same approach.

Now consider the code of Fig. 1.17. In this case there are two input messages and three output sequences. The generator polynomials are:

$$
\begin{aligned}
g^1_1 &= (11) = 1 + D & g^2_1 &= (01) = D & g^3_1 &= (11) = 1 + D \\
g^1_2 &= (01) = D & g^2_2 &= (10) = 1 & g^3_2 &= (10) = 1
\end{aligned}
\qquad (1.42)
$$

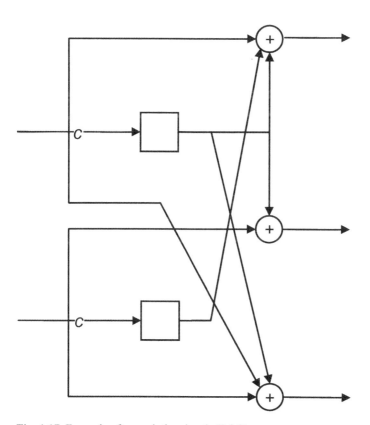

Fig. 1.17. Example of convolutional code (3,2,2)

We have:

$$v^1 = c^1 * g^1_1 + c^2 * g^1_2 \qquad (1.43)$$

$$v^2 = c^1 * g_1^2 + c^2 * g_2^2 \tag{1.44}$$

$$v^3 = c^1 * g_1^3 + c^2 * g_2^3 \tag{1.45}$$

that is

$$v_l^1 = c_l^1 + c_{l-1}^1 + c_{l-1}^2 \tag{1.46}$$

$$v_l^2 = c_l^2 + c_{l-1}^1 \tag{1.47}$$

$$v_l^3 = c_l^1 + c_l^2 + c_{l-1}^1 \tag{1.48}$$

$$v = \left(v_0^1 v_0^2 v_0^3, v_1^1 v_1^2 v_1^3, v_2^1 v_2^2 v_2^3, \cdots \right) \tag{1.49}$$

In general, a code *(n,k,m)* has the generator matrix given in Eq. (1.50)

$$G = \begin{pmatrix} G_0 & G_1 & G_2 & \cdots & G_m & & \cdots \\ 0 & G_0 & G_1 & \cdots & G_{m-1} & G_m & \cdots \\ \vdots & & & & & & \vdots \\ \cdots & & & & & & \cdots \end{pmatrix} \tag{1.50}$$

where G_l is a submatrix *k x m*:

$$G = \begin{pmatrix} g_{1,l}^1 & g_{1,l}^2 & \cdots & g_{1,l}^m \\ g_{2,l}^1 & g_{2,l}^2 & \cdots & g_{2,l}^m \\ \vdots & & & \vdots \\ g_{k,l}^1 & g_{k,l}^2 & \cdots & g_{k,l}^m \end{pmatrix} \tag{1.51}$$

Definition 1.5.14 The *memory of the i-th input* of the encoder is given by

$$m_i = \max_{i \le j \le n} \left\{ \deg g_{i,j}(D) \right\} \tag{1.52}$$

The *encoder memory* is defined as:

$$m = \max_{i \le k} m_i \tag{1.53}$$

The *total memory* is defined as

$$K = \sum_{i=1}^{k} m_i \tag{1.54}$$

Also for convolutional codes there is a systematic form even though, in general, the assumption that for each convolutional code there is an equivalent one in systematic form is not valid. In general the best codes are non-systematic.

A code in non-systematic form needs an inverse function to find the k data bit again. The received message is $v(D) = c(D)G(D)$. In order to recover $c(D)$, the inversion shown in Eq. (1.55) has to be applied.

$$c(D) = v(D) \cdot G^{-1}(D) = c(D) \cdot G(D) \cdot G^{-1}(D) \text{ with } G(D) \cdot G^{-1}(D) = I \tag{1.55}$$

Definition 1.5.15 A code is called *catastrophic* if $G^{-1}(D)$ does not exist: in this case a finite number of errors in the channel produces an infinite number of decoding errors.

Definition 1.5.16 The *minimum free distance* d_{free} is:

$$d_{free} = \min\{d(v, v') : c \neq c'\} \tag{1.56}$$

therefore it is the minimum Hamming distance of two codewords.

The decoding algorithm used in this kind of codes is the algorithm of maximum likelihood. The discoverer of this algorithm is Andrea Viterbi. The state of the encoder is defined as the content of memory cells. If K is the total memory, the number of states is 2^K. The state diagram consists of knots, representing the states, and lines, representing the state transitions. Figure 1.18 shows the state diagram of the code introduced in Fig. 1.16.

For example, if we are in the state S_2 and the input is a 0, we reach the state S_1, meanwhile the output, following Eqs. (1.30) and (1.31), will be 01. On the other hand, if the input is a 1, we can reach the state S_3 and the output will be 10.

A Trellis diagram can be deduced from the state diagram by tracing all the possible input/output sequences and the state transitions (Fig. 1.19).

When the first bit enters the encoder, it is possible to shift to state (10) if the input is a 1 or to state (00) if the input is a 0. The following symbol can reach any of the other states. In Fig. 1.19 we represent the transitions caused by a 0 input with an arrow pointing downward, while the arrows pointing upwards represent transitions caused by a 1 input.

The Viterbi decoding algorithm makes use of the Trellis diagram and describes a procedure of maximum likelihood decoding.

The sequence of the output code can be reached by tracing a path specified by the input sequence. For example, if the input sequence is $c = (11001)$, the output will be $v = (1110101111)$. Starting from this observation, the Viterbi decoding algorithm is now described.

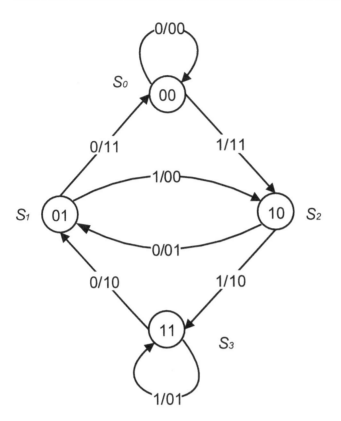

Fig. 1.18. State diagram

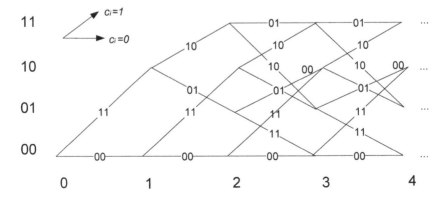

Fig. 1.19. Trellis diagram

Viterbi Algorithm.

- Step 1: since it is not possible to wait for the whole word which could be endless, truncate the word at time *b*.
- Step 2: for each state calculate all the possible paths that arrive in that state.
- Step 3: calculate the free distance between the found word and the received one.
- Step 4: decode the word with the minimum free distance.

Considering the code of Fig. 1.16 once again, suppose we receive 1000011000.... (Fig. 1.20). Choose $b=3$, then truncating the word we get 100001.

– State 00

1. if we have 00 00 00 => 2 errors;
2. if we have 10 01 00 => 3 errors, therefore the path is discarded.

– State 01

1. if we have 00 10 01 => 3 errors, therefore the path is discarded;
2. if we have 10 11 01 => 4 errors, therefore the path is discarded.

In order to conclude the decoding, consider the other two states and then decode the path with the smallest number of errors. Note that more than a single path might admit the minimum number of errors: this is the case of unrecoverable errors and we can choose one of those paths arbitrarily.

The algorithm here introduced is applicable to a binary symmetric channel, where the metric used is the free distance, however, it is extendible by using each type of metric eventually non-symmetric.

The disadvantage in the use of convolutional codes is that a constructive method to generate them does not exist; for values K of practical interest the search for good codes must be carried out in an exhaustive way by trying out all the possible connections, with some rules to reject the bad solutions as fast as possible.

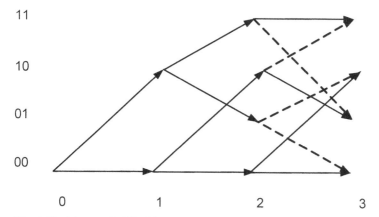

Fig. 1.20. Scheme of a Viterbi algorithm for a convolutional code

Besides, despite the sure advantage of a high hardware parallelism, the decoding requires a lot of memory to store the surviving paths and it is necessary to wait for the end of the transmission to decide on all the information bits. Their greatest use is in concatenated codes and in turbo codes, generally combined with a cyclic code or with a non-symmetric error channel.

1.5.3 Burst type errors

Before choosing a good code for a particular application, it is fundamental to know the type of errors that the code must be able to correct. Till now we have not made any hypothesis on the nature of errors, supposing them to be random and uniformly distributed. There are channels in which all the errors are reunited in packets called bursts: e.g. compact discs, where the error source is the classical scratch, or digital recordings. Another example of this kind of error is the noise during transmission.

Definition 1.5.17 A *burst* of length l is defined as a vector whose components different from 0 are confined in l consecutive positions.

For example, the vector $e = (00001011010000)$ is a burst of length 6.

Definition 1.5.18 Two words have a *burst distance d* if the two words differ of a burst of length d.

Note that the distance just defined is not really a distance in the classical sense because the triangular inequality is not valid.

Similarly to what said for block codes, a codeword is seen as a point in the space. For each word we can build a circle of radius l. The code corrects burst of length l if such circles are separated.

Definition 1.5.19 A linear code is said to detect burst-errors of length l if there are not any burst-error vectors of length l that, added to a codeword, produce another codeword.

Theorem 1.5.2 If a linear code $[n,k]$ detects all the bursts of length l, it must have l parity symbols: $n - k \geq l$ at least.

Theorem 1.5.3 *Reiger Inequality*: if a linear code $[n,k]$ corrects all the bursts of length l, it must have at least $2l$ parity symbols: $n - k \geq 2l$.

The last theorem means that, given n and k, the correction capability of burst-errors is

$$l \leq \left[\frac{n-k}{2} \right] \qquad (1.57)$$

where the square brackets indicate the floor function.

Definition 1.5.20 A code is called *optimum* if $n = 2^{n-k-l+1} - 1$.

Generally speaking, what we should do in order to correct burst errors is to use already known codes combined with interlacing techniques. These techniques change the order in which the digits are transmitted.

We have so far considered messages $m_1, m_2, \ldots$ encoded in the codewords c_1, $c_2, \ldots$ transmitted in a sequential way. Interlacing $c_1, c_2, \ldots$ with depth s means that for $i = 0, 1, 2, \ldots$ the digits of the codewords are transmitted in the order:

$$c_{is+1,1}, c_{is+2,1}, \cdots, c_{is+s,1}, c_{is+1,2}, c_{is+2,2}, \cdots, c_{is+1,n}, \cdots, c_{is+s,n} \qquad (1.58)$$

Conceptually, this is analogous to reorder the bits in a matrix, listing the words $c_{is+1}, \ldots, c_{is+s}$ in rows and transmitting the bits in columns:

$$
\begin{array}{ccccc}
c_{is+1,1} & c_{is+1,2} & c_{is+1,3} & \cdots & c_{is+1,n} \\
c_{is+2,1} & c_{is+2,2} & c_{is+2,3} & \cdots & c_{is+2,n} \\
\vdots & & & & \vdots \\
c_{is+s,1} & c_{is+s,2} & c_{is+s,3} & \cdots & c_{is+s,n}
\end{array} \qquad (1.59)
$$

Suppose we have a linear code with generator matrix:

$$
G = \begin{pmatrix}
1 & 0 & 0 & 1 & 1 & 0 \\
0 & 1 & 0 & 1 & 0 & 1 \\
0 & 0 & 1 & 0 & 1 & 1
\end{pmatrix} \qquad (1.60)
$$

The codewords c_1, c_2, c_3, c_4, c_5 and c_6 to be transmitted are:

$$
\begin{aligned}
c_1 &= 100110 & c_4 &= 010101 \\
c_2 &= 010101 & c_5 &= 100110 & \qquad (1.61) \\
c_3 &= 111000 & c_6 &= 111000
\end{aligned}
$$

Without interleaving they would be sent serially as
100110010101111000010101100110111000.

If these codewords are interlaced with depth 3, then the first digits of c_1, c_2, c_3 that is 1, 0, 1 are transmitted first and the sent message will be:
101011001110100010.

Note that, in this string, the digits of c_1 appear in the positions $3i + 1$ with $0 \leq i \leq 5$. Subsequently, the reordering of c_4, c_5 and c_6 produces 011101001110010100.

The interest in such techniques lies in the fact that errors involving consecutive symbols of an interlaced code of C will corrupt different codewords, provided that we choose a suitable t by observing the error statistics on the channel. Besides, the correction capability of an interlaced code increases remarkably.

Theorem 1.5.4 Let C be a correction code of burst of length l. If C is interlaced with depth s then all the bursts of length sl at the most will be corrected.

The interlacing of depth s has the disadvantage that s codewords must be encoded before each of them is transmitted. An interlace technique with delay r is therefore often used. Using s words of C, we build a matrix where the first row is composed of all the first symbols of the s words of C; the second row is composed of all the second symbols of the s words of C delayed of r positions toward the right of the preceding row and so forth. We get the matrix:

$$
\begin{array}{ccccccccc}
c_{1,1} & c_{2,1} & \cdots & c_{s+1,1} & c_{s+2,1} & \cdots & c_{2s+1,1} & c_{2s+2,1} & \cdots & c_{(n-1)s+1,1} \\
* & * & \cdots & c_{1,2} & c_{2,2} & \cdots & c_{s+1,2} & c_{s+2,2} & \cdots & c_{(n-2)s+1,2} \\
\vdots & & & * & * & & c_{1,3} & c_{2,3} & \cdots & c_{(n-3)s+1,3} \\
\vdots & & & \vdots & \vdots & & \vdots & \vdots & & \vdots \\
* & * & \cdots & * & * & \cdots & * & * & \cdots & c_{1,n}
\end{array}
\tag{1.62}
$$

The columns of the matrix are sent out subsequently, at the beginning of the process the stars are replaced by 0s. Consider the 6 codewords of the preceding example. If we use an interlace with delay 1 we get:

$$
\begin{array}{ccccccccccc}
1 & 0 & 1 & 0 & 1 & 1 & \cdots \\
* & 0 & 1 & 1 & 1 & 0 & 1 & \cdots \\
* & * & 0 & 0 & 1 & 0 & 0 & 1 & \cdots \\
* & * & * & 1 & 1 & 0 & 1 & 1 & 0 & \cdots \\
* & * & * & * & 1 & 0 & 0 & 0 & 1 & 0 & \cdots \\
* & * & * & * & * & 0 & 1 & 0 & 1 & 0 & 0 & \cdots
\end{array}
\tag{1.63}
$$

Theorem 1.5.5 Let C be a correction code of burst-errors of length l. If C is interlaced with depth s and delay n, then all the bursts of length $l(sn+1)$ will be corrected.

In practical applications, the concatenation of two codes is often used. Let C_1 be a linear code $[n_1, k_1, d_1]$ and C_2 a linear code $[n_2, k_2, d_2]$. The cross-interlacing of C_1 with C_2 is done as follows: the messages are first encoded using C_1 and the resulting codewords are interlaced with depth k_2. Now the columns are seen as messages and encoded using C_2. The resulting words can be interlaced with depth s with or without delay.

The advantage consists in the detection capability of C_2 to recognize $(d_2 - 1)$ errors. If errors are detected in a codeword of C_2, then all the digits of the codeword are flagged and treated as digits that could be corrupted. Subsequently, the codewords of C_1 are considered. If we know that $(n_1 - d_1 + 1)$ digits in a codeword c of C_1 are correct, then we can always find the $(d_1 - 1)$ digit.

For example, let C_1 and C_2 be two codes with generator matrices:

$$G_1 = \begin{pmatrix} 1 & 0 & 0 & 0 & 1 & 1 & 1 & 0 \\ 0 & 1 & 0 & 0 & 1 & 1 & 0 & 1 \\ 0 & 0 & 1 & 0 & 1 & 0 & 1 & 1 \\ 0 & 0 & 0 & 1 & 0 & 1 & 1 & 1 \end{pmatrix} \tag{1.64}$$

$$G_2 = \begin{pmatrix} 1 & 0 & 0 & 1 & 1 & 0 \\ 0 & 1 & 0 & 1 & 0 & 1 \\ 0 & 0 & 1 & 0 & 1 & 1 \end{pmatrix} \tag{1.65}$$

Messages to be encoded are: $m_1 = 1000$, $m_2 = 1100$ and $m_3 = 1010$ with cross-interlacing of C_1 with C_2. C_2 is interlaced with depth $s = 3 = d_1 - 1$. The encoding with C_1 gives:

$$c_1 = m_1 G_1 = 10001110 \tag{1.66}$$

$$c_2 = m_2 G_1 = 11000011 \tag{1.67}$$

$$c_3 = m_3 G_1 = 10100101 \tag{1.68}$$

By interlacing these codewords with depth $k_2=3$, we get the following messages: 111, 010, 001, 000, 100, 101, 110, 011. Now these messages are encoded by using C_2 and we get 8 codewords that must be interlaced with depth $s = 3$:

$$\begin{array}{ll} c_1' = 111000 & c_5' = 100110 \\ c_2' = 010101 & c_6' = 101101 \\ c_3' = 001011 & c_7' = 110011 \\ c_4' = 000000 & c_8' = 011110 \end{array} \tag{1.69}$$

The first string to be written is 100110101010001011011000001011010001.

Suppose we have errors in the first 6 bits and we read the word 011001101010001011. In order to decode the message, we have to apply an inverse procedure with respect to the one described above. First of all, the message has to be de-interlaced with depth $s=3$, obtaining: 001000, 100101, 111011.

C_2 recognizes the errors in all the 3 codewords, hence all the 18 digits are marked. Suppose that no other errors occurred and therefore there will be no other marked digits. Removing the effect of the interlace of depth k_2, we get:

$$c_1 = ***01110 \tag{1.70}$$

$$c_2 = ***00011 \tag{1.71}$$

$$c_3 = ***00101 \qquad\qquad (1.72)$$

There is just one way of replacing each marked digit with 0 or 1 to produce codewords and the codewords produced are exactly c_1, c_2, c_3.

Obviously, there are a lot of interlacing techniques that change the order of the bits in the written message. The choice of the suitable interlacing technique is a very important choice since a bad interlace could even be worse than the absence of a correction code.

Bibliography

A. Abedi, A. K. Khandani, "An Analytical Method for Approximate Performance Evaluation of Binary Linear Block Codes" in *IEEE Transactions on Communications*, Vol. 52, February 2004.

E. F. Assmus, J. D. Key, "Design and Their codes", *Cambridge University Press*, 1992.

R. E. Blahut, "Theory and Practice of Error Control Codes", *Addsion-Wesley Publishing Company*, 1983.

A. R. Calderbank, "The Art of Signaling: Fifty Years of Coding Theory" in *IEEE Transactions on Information Theory*, Vol. 44, October 1998.

D. J. Costello, J. Hagenauer, H. Imai, S. B. Wicker, "Applications of Error-Control Coding" in *IEEE Transactions on Information Theory*, Vol. 44, October 1998.

D. J. Costello, J. Hagenauer, H. Imai, S. B. Wicker, "Applications of Error-Control Coding" in *IEEE Transactions on Information Theory*, Vol. 6, October 1998.

D. R. Douglas, "Cryptography: Theory and Practice", *CRC Press*, 1995.

W. Feng, J. Yuan, B. Vucetic, "A Code-Matched Interleaver Design for Turbo Codes" in *IEEE Transactions on Communications*, Vol. 50, June 2002.

F. Guida, E. Montolivo, G. M. Pascetti, "Evaluation of the Performance of Binary Error Correcting Codes over Gaussian QAM Channels" in *Electronics Letters*, Vol. 30, November 1994.

D. R. Hankerson, D. G. Hoffman, D. A. Leonard, C. C. Lindner, K. T. Phelps, C. A. Rodger, J. R. Wall, "Coding Theory and Cryptography the essentials", *Marcel Dekker Inc.*, 1991.

R. Hill, "A First Course in Coding Theory", in *Claredon Press*, Oxford 1986.

A. Jiri, "Foundation of Coding", *John Wiley and Sons Inc.*, 1991.

G. A. Jones, J. M. Jones, "Informations and Coding Theory", *Springer*, 2000.

A. S. A. G. Khaled, R. J. McEliece, A. M. Odlyzka, H. C. A. von Tilborg, "On the Existence of Optimum Cyclic Burst-Correcting Codes" in *IEEE Transactions on Information Theory*, Vol. 32, November 1986.

S. Lin, D. J. Costello, "Error Control Coding: Fundamentals and Applications", *F. F. Kuo Editor*, 1983.

F. J. Macwilliams, N. J. A. Sloane, "The Theory of Error-Correcting Codes", *North-Holland Publishing Company*, 1978.

H. B. Mann, "Error Correcting Codes", *John Wiley & Sons Inc.*, 1968.

B. Marcus, J. Rosenthal, "Codes, Systems and Graphical Models", *Springer-Verlag*, 2001.

T. K. Moon, "Error Correcting Coding – Mathematical Methods and Algorithms", *John Wiley and Sons Inc.*, 2005.

R. H. Morelos-Zaragoza, "The Art of Error-Correcting Coding", *John Wiley and Sons Inc.*, 2002.

C. Nill, C. E. W. Sundberg, "List and Soft Symbol Output Viterbi Algorithms: Extensions and Comparisons" in *IEEE Transactions on Communications*, Vol. 2/3/4, February/March/April 1995.

O. Papini, "Algebre discrete et codes correcteurs", *Springer-Verlag*, 1995.

W. W. Peterson, E. J. Weldon, "Error-Correcting Codes", *MIT Press*, 1998.

V. Pless, "Introduction to the Theory of Error-Correcting Codes", *John Wiley and Sons*, 1998.

M. Purser, "Introduction to Error-Correcting Codes", *Artech House*, 1995.

S. Roman, "Introduction to Coding and Information Theory", *Springer-Verlag*, 1997.

C. E. Shannon, "A Mathematical Theory of Communication" in *Bell System Tech. J.*, 1948.

H. C. A. van Tilborg, "Coding Theory and Applications", *Springer-Verlag*, 1989.

J. H. von Lint, "Introduction to Coding Theory", *Springer-Verlag*, 1999.

S. B. Wicker, "Error Control for Digital Communication and Storage", *Prentice Hall*, 1995.

J. Yuan, B. Vucetic, W. Feng, "Combined Turbo Codes and Interleaver Design" in *IEEE Transactions on Communications*, Vol. 47, April 1999.

2 Error correction codes

2.1 Hamming codes

The Hamming code, conceived by Richard Hamming at Bell laboratories in 1950, is the first error-correcting code to be used in applications. A Hamming code is substantially a perfect linear code able to correct a single error.

The codeword length is 2^s-1 of which s bits are the parity ones.

The encoding and decoding operations for such code are illustrated in Fig. 2.1.

It has been said (Sect. 1.5.1) that, for a linear code, the encoding and decoding procedures can be described through matrix operations.

Let m be the data vector of length k to be encoded and G the generator matrix of the code. The codeword c of length n is obtained as:

$$c = m \cdot G \tag{2.1}$$

The description of the encoding and decoding operations will proceed through the example of a Hamming $C[7,4]$ code.

For a binary Hamming code $C[7,4]$ the generator matrix could be:

$$G = \begin{pmatrix} 1 & 0 & 0 & 0 & 0 & 1 & 1 \\ 0 & 1 & 0 & 0 & 1 & 0 & 1 \\ 0 & 0 & 1 & 0 & 1 & 1 & 0 \\ 0 & 0 & 0 & 1 & 1 & 1 & 1 \end{pmatrix} \tag{2.2}$$

The vector $m=(0110)$ is encoded into c as:

$$c = mG = (0110) \begin{pmatrix} 1 & 0 & 0 & 0 & 0 & 1 & 1 \\ 0 & 1 & 0 & 0 & 1 & 0 & 1 \\ 0 & 0 & 1 & 0 & 1 & 1 & 0 \\ 0 & 0 & 0 & 1 & 1 & 1 & 1 \end{pmatrix} = (0110011) \tag{2.3}$$

Let e be the error vector, the read vector r is given by:

$$r = c + e \tag{2.4}$$

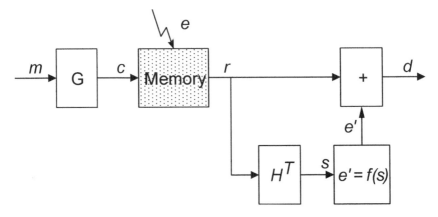

Fig. 2.1. Representation of the encoding and decoding operations for a Hamming code

In the decoding phase, the syndrome is obtained as

$$s = rH^T, \tag{2.5}$$

where H, the *parity matrix*, is constructed to satisfy the following equality

$$GH^T = 0. \tag{2.6}$$

A possible matrix H (remember that an infinite number of equivalent ones exists) is

$$H = \begin{pmatrix} 0 & 0 & 0 & 1 & 1 & 1 & 1 \\ 0 & 1 & 1 & 0 & 0 & 1 & 1 \\ 1 & 0 & 1 & 0 & 1 & 0 & 1 \end{pmatrix}. \tag{2.7}$$

As already said, the syndrome for the vector c (Eq. (2.3)) is null.

$$s = cH^T = (0110011) \cdot \begin{pmatrix} 0 & 0 & 0 & 1 & 1 & 1 & 1 \\ 0 & 1 & 1 & 0 & 0 & 1 & 1 \\ 1 & 0 & 1 & 0 & 1 & 0 & 1 \end{pmatrix}^T = (0 \quad 0 \quad 0) \tag{2.8}$$

Taking Eq. (2.5) again and replacing r we get

$$s = rH^T = (c + e) \cdot H^T = mGH^T + eH^T = eH^T \tag{2.9}$$

therefore the syndrome is univocally determined by the error vector e.

The i-th column of H represents the syndrome for an error in the i-th position of the vector r.

A Hamming code is completely described by the matrix H, which defines all the characteristics of the code such as the length n and the number of parity bits $n-k$. There exists a constructive method to find a matrix H of a given Hamming code. In the binary case the matrix H for a code C is constituted by the $2^{n-k}-1$ columns that can be written with $n-k$ bits, with the following constraints:

- all the columns must be different;
- all the columns must be different from the null vector.

When the matrix H is written in order from 1 to $2^{n-k}-1$, the syndrome value gives the position of the error. Reconsider the matrix H of the previous example and suppose that, for occurrence of an error on the bit underlined, the codeword $c=(0110011)$ is re-read as $r=(01\underline{0}0011)$. The syndrome computation is:

$$s = rH^T = (0100011) \cdot \begin{pmatrix} 0 & 0 & 0 & 1 & 1 & 1 & 1 \\ 0 & 1 & 1 & 0 & 0 & 1 & 1 \\ 1 & 0 & 1 & 0 & 1 & 0 & 1 \end{pmatrix}^T = (0 \quad 1 \quad 1) = 3 \qquad (2.10)$$

The syndrome value gives the position of the bit error.

The matrix G of this example is such that the code results systematic, in other words, it is possible to recognize the message m inside the codeword $c=mG$.

Through permutation of columns or multiplication by a matrix with non-null determinant, it is always possible to get a systematic code. The parity matrix of our example, written in systematic form is H'.

$$H' = \begin{pmatrix} 0 & 1 & 1 & 1 & 1 & 0 & 0 \\ 1 & 0 & 1 & 1 & 0 & 1 & 0 \\ 1 & 1 & 0 & 1 & 0 & 0 & 1 \end{pmatrix} \qquad (2.11)$$

Supposing we receive $r=(01\underline{0}0011)$ we have:

$$s = rH'^T = (0100011) \cdot \begin{pmatrix} 0 & 1 & 1 & 1 & 1 & 0 & 0 \\ 1 & 0 & 1 & 1 & 0 & 1 & 0 \\ 1 & 1 & 0 & 1 & 0 & 0 & 1 \end{pmatrix}^T = (1 \quad 1 \quad 0) \qquad (2.12)$$

The syndrome identifies the error in a univocal way but it is not equal to its position. The error position coincides with the position of the syndrome's value in H.

The definition of the Hamming code as a perfect code fixes the length of the codewords n. In the binary case we have that $n=2^{n-k}-1$ and $k=2^{n-k}-1-(n-k)$.

In the case of a length k' fixed and different from $2^{n-k}-1-(n-k)$, it is always possible to construct a Hamming code by using the shortening operation described in the preceding chapter. Note that, in this case, the code will not result to be perfect anymore.

2.2 Reed-Muller codes

Reed-Muller codes were described by Muller through the use of Boolean functions in 1954. In the same year Reed described a decoding algorithm for codes corrector of more than one error of the Muller type based on a set of parity equations. Since then, these codes have been remembered as Reed-Muller codes.

Reed-Muller codes are very good for short lengths, but they have inferior performance, compared to BCH or Reed-Solomon codes, as the length grows. Today there are many ways to describe them; here we have preferred the original Muller method, by means of Boolean functions.

Definition 2.2.1 A Boolean function in m variables $f(x_1, x_2, ..., x_m)$ is defined as a map from the vector space V_m of binary m-ple $(x_1, x_2, ..., x_m)$ toward the set of binary numbers $\{0, 1\}$.

The Boolean functions are completely described by the truth table containing $m+1$ rows. The first m rows form a matrix $m \times 2^m$ which contains all the binary 2^m m-ple as columns. The last row represents the binary value assigned to each m-pla by the Boolean function.

Consider, for example, a Boolean function f with 4 variables (v_1, v_2, v_3, v_4), described by the truth table of Table 2.1; it is easy to observe that $f = v_4 + v_3 + v_2 + v_1$.

Table 2.1. Example of Boolean function

v_1	0	0	0	0	0	0	0	0	1	1	1	1	1	1	1	1
v_2	0	0	0	0	1	1	1	1	0	0	0	0	1	1	1	1
v_3	0	0	1	1	0	0	1	1	0	0	1	1	0	0	1	1
v_4	0	1	0	1	0	1	0	1	0	1	0	1	0	1	0	1
f	0	1	1	0	1	0	0	1	1	0	0	1	0	1	1	0

Definition 2.2.2 A Reed-Muller code $R(r,m)$, of order r with $0 \leq r \leq m$ and with length $n = 2^m$, is the set of the images of the Boolean functions with m variables of degree $\leq r$.

Suppose we want to construct the generator matrix of a Reed-Muller code $R(2,4)$. According to the definition, the code has length $n = 2^4$ and is formed by the images of the functions of second degree at most.

In the binary field the multiplication of two elements is computed as a logical AND operation between them. This is the reason why, for Boolean functions, the multiplication between two variables v_1 and v_2 consists in the logical AND operation bit by bit between the coefficients of the multiplicands.

For what said, note that there is no need to consider squares of individual variables since v_i and v_i^2 represent the same Boolean function.

The rows of the generator matrix G are summarized in Table 2.2.

Table 2.2. Generator matrix of $R(2,4)$

1	1	1	1	1	1	1	1	1	1	1	1	1	1	1	1
v_4	0	0	0	0	0	0	0	1	1	1	1	1	1	1	1
v_3	0	0	0	1	1	1	1	0	0	0	0	1	1	1	1
v_2	0	1	1	0	0	1	1	0	0	1	1	0	0	1	1
v_1	1	0	1	0	1	0	1	0	1	0	1	0	1	0	1
v_3v_4	0	0	0	0	0	0	0	0	0	0	0	1	1	1	1
v_2v_4	0	0	0	0	0	0	0	0	0	1	1	0	0	1	1
v_1v_4	0	0	0	0	0	0	0	0	1	0	1	0	1	0	1
v_2v_3	0	0	0	0	0	1	1	0	0	0	0	0	0	1	1
v_1v_3	0	0	0	0	1	0	1	0	0	0	0	0	1	0	1
v_1v_2	0	0	1	0	0	0	1	0	0	0	1	0	0	0	1

The code $R(2,4)$ is a code $C[16,11]$ with minimum distance 4.

Each codeword, as for all block codes, (Sect. 1.5.1), is a linear combination of the rows of the generator matrix.

Theorem 2.2.1 The *dimension k*, i.e. the number of information bits in a code-word, of a Reed-Muller code $R(r,m)$ is

$$k = 1 + \binom{m}{1} + \binom{m}{2} + \cdots + \binom{m}{r} \tag{2.13}$$

Theorem 2.2.2 A Reed-Muller code $R(r,m)$ has minimum distance 2^{m-r}.

The best advantage of Reed-Muller codes is their decoding algorithm, which represents the first non-banal example of majority or threshold decoding. In general, the majority decoding is fast, but not optimal. Instead, for Reed-Muller codes, the majority decoding has the same performances of maximum likelihood decoding.

The decoding algorithm will be illustrated applied to the particular case of a code $R(1,3)$. The generator matrix of this code is

$$G = \begin{pmatrix} 1 & 1 & 1 & 1 & 1 & 1 & 1 & 1 \\ 0 & 0 & 0 & 0 & 1 & 1 & 1 & 1 \\ 0 & 0 & 1 & 1 & 0 & 0 & 1 & 1 \\ 0 & 1 & 0 & 1 & 0 & 1 & 0 & 1 \end{pmatrix} \tag{2.14}$$

The code written in this form is not systematic, thus, if the word to be encoded is $(a_0,\ a_1,\ a_2,\ a_3)$, the word encoded will be $(b_0,\ b_1,\ b_2,\ b_3,\ b_4,\ b_5,\ b_6,\ b_7) = a_0 1 + a_1 v_1 + a_2 v_2 + a_3 v_3$ where the b_i are calculated with the following equations:

$$b_0 = a_0$$
$$b_1 = a_0 + a_3$$
$$b_2 = a_0 + a_2$$
$$b_3 = a_0 + a_2 + a_3$$
$$b_4 = a_0 + a_1$$
$$b_5 = a_0 + a_1 + a_3$$
$$b_6 = a_0 + a_1 + a_2$$
$$b_7 = a_0 + a_1 + a_2 + a_3$$

(2.15)

Working out Eqs. (2.15), it is easy to verify the following sets of equations:

$$a_1 = b_0 + b_4$$
$$a_1 = b_2 + b_6$$
$$a_1 = b_3 + b_7$$
$$a_1 = b_1 + b_5$$

(2.16)

$$a_2 = b_0 + b_2$$
$$a_2 = b_1 + b_3$$
$$a_2 = b_4 + b_6$$
$$a_2 = b_5 + b_7$$

(2.17)

$$a_3 = b_0 + b_1$$
$$a_3 = b_2 + b_3$$
$$a_3 = b_4 + b_5$$
$$a_3 = b_6 + b_7$$

(2.18)

$$a_0 = b_0$$
$$a_0 = b_1 + b_6 + b_7$$
$$a_0 = b_2 + b_5 + b_7$$
$$a_0 = b_3 + b_4 + b_7$$

(2.19)

The majority decoding is based on the majority estimate of these coefficients. Assuming the message received or read is $r=(r_0, r_1, r_2, r_3, r_4, r_5, r_6, r_7)$, we calculate the following sums:

$$a_1^1 = r_0 + r_4$$
$$a_1^2 = r_2 + r_6$$
$$a_1^3 = r_3 + r_7$$
$$a_1^4 = r_1 + r_5$$

(2.20)

The bit a_1 is decoded as

$$a_1 = maj\{a_1^1, a_1^2, a_1^3, a_1^4\}$$

(2.21)

In case there are no errors, each of the four Eqs. (2.20) gives the same result, while a single error causes exactly one of the expressions to be incorrect: in this case, using the majority decoding, we are able to correct the error, since all the other expressions will give the correct result. If 2 errors occur there are two erroneous equations but it is not possible to use a majority decoding because we can not decide which equations' pair is the correct one.

After having found a_1, the decoding of the other bits proceeds in a similar way by computing the following equations and taking majority decisions:

$$a_2^1 = r_0 + r_2$$
$$a_2^2 = r_1 + r_3$$
$$a_2^3 = r_4 + r_6$$
$$a_2^4 = r_5 + r_7$$

(2.22)

$$a_2 = maj\{a_2^1, a_2^2, a_2^3, a_2^4\}$$

(2.23)

$$a_3^1 = r_0 + r_1$$
$$a_3^2 = r_2 + r_3$$
$$a_3^3 = r_4 + r_5$$
$$a_3^4 = r_6 + r_7$$

(2.24)

$$a_3 = maj\{a_3^1, a_3^2, a_3^3, a_3^4\}$$

(2.25)

At this point there is only one bit left to decode. In order to find it, the already decoded coefficients are removed from the received vector:

$$r' = r - (a_1, a_2, a_3) \cdot (v_1, v_2, v_3)$$

(2.26)

The value of r' provides a majority estimate of the bit a_0.

Suppose we have the message $m=(1101)$ to encode. The encoded message c is calculated as:

$$c = mG = v_0 + v_1 + v_3 = 10100101 \tag{2.27}$$

Suppose that an error occurs and we therefore receive $r=(101\underline{1}0101)$ with one error on the bit underlined.

Equations (2.20), (2.22) and (2.24) give:

$$
\begin{aligned}
a_1^1 &= r_0 + r_4 = 1 \\
a_1^2 &= r_2 + r_6 = 1 \\
a_1^3 &= r_3 + r_7 = 0 \\
a_1^4 &= r_1 + r_5 = 1
\end{aligned}
\tag{2.28}
$$

By majority we have $a_1=1$.

$$
\begin{aligned}
a_2^1 &= r_0 + r_2 = 0 \\
a_2^2 &= r_1 + r_3 = 1 \\
a_2^3 &= r_4 + r_6 = 0 \\
a_2^4 &= r_5 + r_7 = 0
\end{aligned}
\tag{2.29}
$$

By majority we have $a_2=0$.

$$
\begin{aligned}
a_3^1 &= r_0 + r_1 = 1 \\
a_3^2 &= r_2 + r_3 = 0 \\
a_3^3 &= r_4 + r_5 = 1 \\
a_3^4 &= r_6 + r_7 = 1
\end{aligned}
\tag{2.30}
$$

By majority we have $a_3=1$.
To decode a_0 we calculate r', using Eq. (2.26).

$$
\begin{aligned}
r' &= r - (a_1, a_2, a_3) \cdot (v_1, v_2, v_3) \\
&= (10110101) - (01011010) = (11101111)
\end{aligned}
\tag{2.31}
$$

By majority $a_0=1$ and the erroneous bit is the fourth: the message is correctly decoded as 1101.

2.3 Cyclic codes

Cyclic codes are the block linear codes most used in applications due to their regular algebraic structure. In the following we will recall the fundamental notions of algebra required for their understanding; then we will use those notions to define them and, finally, we will go through a detailed description of the two most important classes of cyclic codes, BCH and Reed-Solomon codes.

2.3.1 Introduction

2.3.1.1 Algebraic notions

Definition 2.3.1 A *group G* is a set of elements, where an operation * is defined so that:

- G is closed for *, that is if $g \in G$ and $h \in G \Rightarrow g*h \in G$;
- * is associative;
- G has the identity, that is there is an element $e \in G$ such that $e*g=g*e=g$ for each element g of the group;
- each element of the group has an inverse, that is for every g there is a g^{-1} in G such that $g^{-1} * g = g * g^{-1} = e$.

Definition 2.3.2 The *order of a group* is the number of its elements.

We shall now consider only finite groups.

Definition 2.3.3 The *order of an element* $g \in G$ is the smallest natural integer $r \neq 0$ such that $g^r=e$.

Definition 2.3.4 A *field F* is a set of elements, where two operations + and * are defined so that:

- F is a commutative group for the operation +, the identity element is indicated with 0;
- $F \setminus \{0\}$ is a commutative group for the *, the identity element is indicated with 1;
- the distributive property applies, i.e. $a*(b+c)=(a*b)+(a*c)$.

The fields of finite order are particularly interesting for the coding theory. They were first discovered by Evariste Galois and this is the reason why they are called Galois fields and indicated as *GF(q)*.

Theorem 2.3.1 The integers $\{0,1,2,...,p-1\}$ with p prime form a field *GF(p)* with the operations of sum *mod p* and product *mod p*.

For example, for $p=3$ we have the field $GF(3)=\{0,1,2\}$ where the operations of sum and product are described by Tables 2.3 and 2.4 respectively.

Table 2.3. Sum in GF(3)

+	0	1	2
0	0	1	2
1	1	2	0
2	2	0	1

Table 2.4. Product in GF(3)

x	0	1	2
0	0	0	0
1	0	1	2
2	0	2	1

Applying Definition 2.3.3 to Galois field, we can say that, given an element β of the field $GF(q)$, the order of β is the smallest natural number m, such that $\beta^m=1$.

Definition 2.3.5 Let α be an element of $GF(q)$, α is called *primitive element* if it has order q-1.

Theorem 2.3.2 In $GF(q)$ there is a primitive element α. Every non null element of $GF(q)$ can be expressed as power of α, in other words, the multiplicative group $GF(q)$ is cyclic.

Similarly to what said for the multiplication group, for the addition it is possible to give the following definitions.

Definition 2.3.6 The *characteristic* of a Galois field $GF(q)$ is the smallest natural number m such that the sum of m 1s is equal to 0.

From the definition of Galois fields it follows that their characteristic is always a prime number.

We will now indicate with $GF(q)[x]$ the set of polynomials of any degree with coefficients in $GF(q)$.

Definition 2.3.7 A polynomial $f(x)$ is called *monic* if its highest degree coefficient is 1.

Definition 2.3.8 A polynomial $f(x)$ is *irreducible* in $GF(q)$ if it cannot be factorized into a product of polynomials of lower degree in $GF(q)[x]$.

Definition 2.3.9 An irreducible polynomial $p(x)$ in $GF(p)[x]$ of degree m is called *primitive* if the smallest natural number n, such that x^n-1 is a multiple of $p(x)$, is $n=p^m$-1.

It is also possible to demonstrate that the roots $\{\alpha_j\}$ of a primitive polynomial $p(x)$ of degree m in $GF(p)[x]$ have order p^m-1.

Knowing that α has order p^m-1, the p^m-1 consecutive powers of α form a multiplicative group of order p^m-1. From the exponential representation it is easy to recover the polynomial representation by reducing the sequence of powers of α with the operation *modulo p(x)*.

In fact, if $p(x)=x^m+a_{m-1}x^{m-1}+...+a_1x+a_0$ is a primitive polynomial in $GF(p)[x]$ and α is one of its roots, we have $p(\alpha)=\alpha^m+a_{m-1}\alpha^{m-1}+...+a_1\alpha+a_0=0$.

Accordingly:

$$\alpha^m = -a_0 - a_1\alpha - a_2\alpha^2 - ... - a_{m-1}\alpha^{m-1} \tag{2.32}$$

This means that each power of α of degree greater than or equal to m can be expressed as polynomials of degree $m-1$ at most.

There are exactly p^m-1 of these polynomials which, added to the null element, form an additive group.

The set of these elements forms a Galois field indicated with $GF(p^m)$. Once again remember that p has to be a prime number.

As an example, we will follow the construction of the field $GF(2^3)$.

Let $p(x)=x^3+x+1$ be a primitive polynomial in $GF(2)[x]$ and α one of its roots. We have the equality $\alpha^3=\alpha+1$ with which it is possible to construct Table 2.5.

Table 2.5. Possible representations of the elements of GF(8)

Exponential representation	Polynomial representation	Binary representation	Decimal representation
0	0	000	0
α^0	1	100	1
α^1	α	010	2
α^2	α^2	001	4
α^3	$\alpha+1$	110	3
α^4	$\alpha^2+\alpha$	011	6
α^5	$\alpha^2+\alpha+1$	111	7
α^6	α^2+1	101	5

The sum between field elements is performed by summing *mod p* the coefficients of the polynomial representation.

Suppose we want to sum α^2 and α^5, we have:

$$\alpha^2 + \alpha^5 = \alpha^2 + \alpha^2 + \alpha + 1 = \alpha + 1 = \alpha^3 \tag{2.33}$$

On the contrary, the multiplication is performed by using the exponential representation, summing the exponents *mod (p^m-1)*. The multiplication can also be seen as the product of the polynomial representation *mod p(x)*.

Since the polynomial representation of a finite field $GF(p^m)$ has coefficients in $GF(p)$, $GF(p^m)$ can be seen as a vectorial space over $GF(p)$, as represented by the third column of Table 2.5.

The last column of Table 2.5. gives the decimal form. This form is useful in calculations but it is necessary to underline that it does not represent an equivalence with the decimal numbers; it is an integer that represents a vector.

Finally, this last definition allows us to link the field $GF(p^m)$ to the field $GF(p)$.

Definition 2.3.10 Let α be an element of the field $GF(p^m)$. The *minimal polynomial* of α in $GF(p)$ is the smallest-degree nonzero polynomial $p(x)$ in $GF(p)[x]$ such that $p(\alpha)=0$.

2.3.1.2 Generalities on cyclic codes

Cyclic codes are based on the algebraic notions previously defined. They were first discussed by Prange, between 1957 and 1959, with a series of technical reports. Today they are, perhaps, the most used codes in applications, since they can be implemented by using high-speed shift-register encoders and decoders. We will now deal only with linear cyclic codes, as the non-linearity implies less structure and therefore fewer applications.

Definition 2.3.11 A linear code $C[n,k]$ is called *cyclic* if $(x_1,x_2,...,x_n) \in C =>$ $(x_n,x_1,...,x_{n-1}) \in C$.

Considering cyclic codes, it is very useful to write the vector $a(x)=(a_0,...,a_{n-1})$ as the polynomial $a_0+a_1x+a_2x^2+...+a_{n-1}x^{n-1}$. According to the previous definition, if $a(x) \in C$, then also the right shift $\in C$.

$$a'(x)= (a_{n-1},a_0,...,a_{n-2})= a_{n-1}+a_0x+...+a_{n-2}x^{n-1}$$
$$= x \cdot a(x)\operatorname{mod}(x^n -1) \tag{2.34}$$

We are therefore working in the congruence classes of the polynomials in $GF(q)[x]$ (mod (x^n-1)), which are indicated as $GF(q)[x]/(x^n-1)$. Consider these classes as the set of the polynomials of degree less than n. The polynomials are added and subtracted like in $GF(q)$, but the multiplication is performed modulo (x^n-1).

The following theorem applies:

Theorem 2.3.3 Let C be a linear cyclic code $[n,k]$

1. Within the set of code polynomials in C there is a unique monic polynomial $g(x)$ with minimal degree $r < n$: $g(x)$ is called the *generator polynomial* of C.
2. Every code polynomial $c(x)$ in C can be expressed uniquely as $c(x)=m(x)g(x)$, where $g(x)$ is the generator polynomial of C and $m(x)$ is a polynomial of degree less than $n-r$ in $GF(q)[x]$.
3. The generator polynomial $g(x)$ of C is a factor of x^n-1 in $GF(q)[x]$.

Accordingly, the search of all the divisors of x^n-1 allows us to find all the cyclic codes of length n.

For example the factorization of x^7-1 in irreducible divisors in $GF(2)[x]$ is:

$$x^7 -1 = (x-1)\cdot(x^3 +x +1)\cdot(x^3 +x^2 +1) \tag{2.35}$$

Hence the codes of length 7 over $GF(2)[x]$ different from $\{0\}$ are:

$$C_0 : g_0(x) = x^7 - 1 \tag{2.36}$$

$$C_1 : g_1(x) = x - 1 \tag{2.37}$$

$$C_2 : g_2(x) = x^3 + x + 1 \tag{2.38}$$

$$C_3 : g_3(x) = x^3 + x^2 + 1 \tag{2.39}$$

$$C_4 : g_4(x) = (x - 1) \cdot (x^3 + x + 1) \tag{2.40}$$

$$C_5 : g_5(x) = (x - 1) \cdot (x^3 + x^2 + 1) \tag{2.41}$$

$$C_6 : g_6(x) = (x^3 + x + 1) \cdot (x^3 + x^2 + 1) \tag{2.42}$$

Theorem 2.3.4 If the degree of $g(x)$ is $n{-}k$, then the code C carries k information bits. If $g(x)=g_0+g_1x+...+g_{n-k}x^{n-k}$, then a generator matrix of C is the following:

$$G = \begin{pmatrix} g_0 & g_1 & g_2 & \cdots & g_{n-k} & 0 & 0 & \cdots & 0 \\ 0 & g_0 & g_1 & \cdots & g_{n-k-1} & g_{n-k} & 0 & \cdots & 0 \\ 0 & 0 & g_0 & \cdots & g_{n-k-2} & g_{n-k-1} & g_{n-k} & \cdots & 0 \\ \vdots & & & & & & & & \vdots \\ 0 & 0 & 0 & \cdots & & & & \cdots & g_{n-k} \end{pmatrix} \tag{2.43}$$

Theorem 2.3.5 If $g(x)$ is a polynomial of degree $n{-}k$ and it is a factor of $x^n{-}1$, then $g(x)$ generates a cyclic code $[n,k]$.

As an example we construct the previous code C_2 (Eq. (2.38)) of length 7, with $g(x)=1+x+x^3$. The generator matrix is:

$$G_2 = \begin{pmatrix} 1 & 1 & 0 & 1 & 0 & 0 & 0 \\ 0 & 1 & 1 & 0 & 1 & 0 & 0 \\ 0 & 0 & 1 & 1 & 0 & 1 & 0 \\ 0 & 0 & 0 & 1 & 1 & 0 & 1 \end{pmatrix} \tag{2.44}$$

Theorem 2.3.6 If $g(x)h(x)=x^n{-}1$ in $GF(q)[x]$ and $g(x)$ is the generator polynomial of a code C and $h(x)=h_0+h_1x+...+h_kx$, then the parity matrix H of C is:

$$H = \begin{pmatrix} h_k & h_{k-1} & \cdots & h_0 & 0 & 0 & \cdots & 0 \\ 0 & h_k & \cdots & h_1 & h_0 & 0 & \cdots & 0 \\ \vdots & & & & & & & \vdots \\ 0 & 0 & \cdots & & & & \cdots & h_0 \end{pmatrix} \qquad (2.45)$$

Taking the code C_2 (Eq. (2.38)) again, we can find its parity matrix with $h(x)=(1+x)(1+x^2+x^3)=1+x+x^2+x^4$.

$$H = \begin{pmatrix} 1 & 0 & 1 & 1 & 1 & 0 & 0 \\ 0 & 1 & 0 & 1 & 1 & 1 & 0 \\ 0 & 0 & 1 & 0 & 1 & 1 & 1 \end{pmatrix} \qquad (2.46)$$

2.3.2 BCH codes

2.3.2.1 Definitions

BCH codes belong to the most important class of algebraic codes. They were found through independent researches by Hocquenghem in 1959 and by Bose and Ray-Chauduri in 1960.

For BCH codes the minimum distance can be ensured during construction. Generally speaking, in order to know the minimum distance for a linear code with generator polynomial $g(x)$, it is necessary to compute the distance between all the possible codewords. BCH codes, by imposing some constraints on the generator polynomial, are able to ensure a "designed distance".

Definition 2.3.12 Let β be an element of $GF(q^m)$. Let b be a non-negative integer. A BCH code with "designed" distance d is generated by the polynomial $g(x)$ of minimal degree that has $d-1$ consecutive powers of β: β^b, β^{b+1}, ..., β^{b+d-2} as roots. Given Ψ_i the minimal polynomial of β^{b+i} for $0 \leq i < d-1$, $g(x)$ is computed as:

$$g(x) = LCM\{\psi_0(x), \psi_1(x), \dots, \psi_{d-2}(x)\} \qquad (2.47)$$

and the data carried by the code is $k=n-deg(g(x))$.

Since the generator polynomial is constructed by using minimal polynomials in $GF(q)[x]$, the code exists in $GF(q)$.

It is possible to show that the designed d is at least $2t+1$, hence the code is able to correct t errors.

If we assume $b=1$, and β a primitive element of $GF(q^m)$ the code becomes a *narrow-sense* and *primitive* BCH code of length q^m-1 able to correct t errors. We shall now consider primitive BCH codes.

In the construction of a BCH code two fields are involved: $GF(q)$, i.e. the field where there are the codewords and where $g(x)$ has its coefficients, and $GF(q^m)$, where $g(x)$ has its roots. To encode and decode a message it is necessary to work with both these fields.

From the definition of a BCH code corrector of t errors it follows that every codeword has β^b, β^{b+1},..., β^{b+2t-1} as roots. Vice versa, if a polynomial $v(x)=v_0+v_1x+...+v_{n-1}x^{n-1}$ in $GF(q)$ has β^b, β^{b+1}, ..., β^{b+2t-1} as roots, it is then divisible by every minimal polynomials Ψ_i of β^i (for $b \leq i \leq b+2t-1$), by their lowest common multiple and therefore by the generator polynomial $g(x)$: in other words, it is a codeword.

Definition 2.3.13 A n-pla $v=(v_0, v_1, ..., v_{n-1})$ in $GF(q)$ is a codeword if and only if the polynomial $v(x)=v_0+v_1x+...+v_{n-1}x^{n-1}$ has β^b, β^{b+1}, ..., β^{b+2t-1} as roots.

The parity matrix H of such code is:

$$H = \begin{pmatrix} 1 & \beta^b & \beta^{2b} & \cdots & \beta^{(n-1)b} \\ 1 & \beta^{b+1} & \beta^{2(b+1)} & \cdots & \beta^{(n-1)(b+1)} \\ \vdots & & & & \vdots \\ 1 & \beta^{b+2t-2} & \beta^{2(b+2t-2)} & \cdots & \beta^{(n-1)(b+2t-2)} \\ 1 & \beta^{b+2t-1} & \beta^{2(b+2t-1)} & \cdots & \beta^{(n-1)(b+2t-1)} \end{pmatrix} \tag{2.48}$$

In the following we will consider the particular case of the primitive narrow-sense binary BCH codes. In this case the generator polynomial $g(x)$ has coefficients in $GF(2)$ and it is the product of the minimal polynomials of β, β^2, β^3, β^{2t} where β is a primitive element in $GF(2^m)$. Recalling Eq. (2.48), a primitive narrow-sense binary BCH code has the matrix H in the form:

$$H = \begin{pmatrix} 1 & \beta & \beta^2 & \cdots & \beta^{n-1} \\ 1 & \beta^2 & \beta^{2(2)} & \cdots & \beta^{2(n-1)} \\ \vdots & & & & \vdots \\ 1 & \beta^{2t-1} & \beta^{2(2t-1)} & \cdots & \beta^{(n-1)(2t-1)} \\ 1 & \beta^{2t} & \beta^{2(2t)} & \cdots & \beta^{(n-1)(2t)} \end{pmatrix} \tag{2.49}$$

It is possible to show that in the binary field the minimal polynomial of β^i coincides with the one of β^{2i}, hence the generator polynomial $g(x)$ results to be the product of the minimal polynomials of only the odd roots and the matrix H becomes:

$$H = \begin{pmatrix} 1 & \beta & \beta^2 & \cdots & \beta^{n-1} \\ 1 & \beta^3 & \beta^{3(2)} & \cdots & \beta^{3(n-1)} \\ \vdots & & & & \vdots \\ 1 & \beta^{2t-3} & \beta^{2(2t-3)} & \cdots & \beta^{(n-1)(2t-3)} \\ 1 & \beta^{2t-1} & \beta^{2(2t-1)} & \cdots & \beta^{(n-1)(2t-1)} \end{pmatrix} \tag{2.50}$$

The number of parity bits for a binary BCH code results to be less than or equal to mt. Generally, this number is equal to mt; it is less only when the minimum distance is greater than the designed distance the code is constructed with.

We want, for example, to construct a binary BCH code of length 15. The polynomial has coefficients in $GF(2)[x]$ and the primitive element is α in $GF(2^4)$.

To construct $g(x)$ we must search for the minimal polynomials of α and α^3 in $GF(2)[x]$:

1. α $\Rightarrow$ $\Psi_1 = 1 + x + x^4$
2. α^3 $\Rightarrow$ $\Psi_3 = 1 + x + x^2 + x^3 + x^4$

Accordingly, $g(x) = (1 + x + x^4)(1 + x + x^2 + x^3 + x^4) = 1 + x^4 + x^6 + x^7 + x^8$, therefore the code requires 8 parity bits.

The parity matrix H is:

$$H = \begin{pmatrix} 1 & \alpha & \alpha^2 & \alpha^3 & \alpha^4 & \alpha^5 & \alpha^6 & \alpha^7 & \alpha^8 & \cdots \\ 1 & \alpha^3 & \alpha^6 & \alpha^9 & \alpha^{12} & \alpha^{15} & \alpha^{18} & \alpha^{21} & \alpha^{24} & \cdots \end{pmatrix}$$

$$\begin{matrix} \cdots & \alpha^9 & \alpha^{10} & \alpha^{11} & \alpha^{12} & \alpha^{13} & \alpha^{14} \\ \cdots & \alpha^{27} & \alpha^{30} & \alpha^{33} & \alpha^{36} & \alpha^{39} & \alpha^{42} \end{matrix}$$

(2.51)

Knowing that $GF(16)$ has $1 + x + x^4$ as primitive polynomial, that is $\alpha^4 = \alpha + 1$, it is possible to write the matrix H in $GF(2)$:

$$H = \begin{pmatrix} 1 & 0 & 0 & 0 & 1 & 0 & 0 & 1 & 1 & 0 & 1 & 0 & 1 & 1 & 1 \\ 0 & 1 & 0 & 0 & 1 & 1 & 0 & 1 & 0 & 1 & 1 & 1 & 1 & 0 & 0 \\ 0 & 0 & 1 & 0 & 0 & 1 & 1 & 0 & 1 & 0 & 1 & 1 & 1 & 1 & 0 \\ 0 & 0 & 0 & 1 & 0 & 0 & 1 & 1 & 0 & 1 & 0 & 1 & 1 & 1 & 1 \\ 1 & 0 & 0 & 0 & 1 & 1 & 0 & 0 & 0 & 1 & 1 & 0 & 0 & 0 & 1 \\ 0 & 0 & 0 & 1 & 1 & 0 & 0 & 0 & 1 & 1 & 0 & 0 & 0 & 1 & 1 \\ 0 & 0 & 1 & 0 & 1 & 0 & 0 & 1 & 0 & 1 & 0 & 0 & 1 & 0 & 1 \\ 0 & 1 & 1 & 1 & 1 & 0 & 1 & 1 & 1 & 1 & 0 & 1 & 1 & 1 & 1 \end{pmatrix}$$

(2.52)

BCH codes are linear so that it is possible to encode and decode messages with the aid of matrices. However, when the length of the code and its correction capability increase, the storage of such matrices may be very onerous. One of the greatest advantages of cyclic codes is the possibility to perform encoding and decoding operations without storing the matrices, but using only the generator polynomial and the minimal polynomials of the code.

In general, the decoding of a BCH code is at least 10 times more complicated than the encoding. In the following we will deal only with binary BCH codes, whose structure is presented in Fig. 2.2.

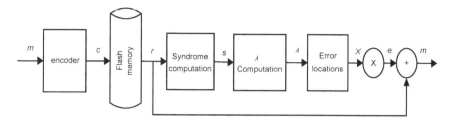

Fig. 2.2. General structure of a binary BCH code

2.3.2.2 Encoding

For all cyclic codes, including BCH, the encoding operations are performed in the same way. For a better understanding of the operations involved, it is useful to consider the message as a polynomial.

Suppose we have a BCH code $[n,k]$ with generator polynomial $g(x)$ and a message $m(x)$ to encode, written as a polynomial of degree $k-1$.

First of all, the message $m(x)$ is multiplied by x^{n-k} and subsequently divided by $g(x)$, obtaining a quotient $q(x)$ and a remainder $r(x)$ in accordance with Eqs. (2.53) and (2.54).

$$\frac{m(x) \cdot x^{n-k}}{g(x)} = q(x) + \frac{r(x)}{g(x)} \tag{2.53}$$

$$m(x) \cdot x^{n-k} + r(x) = q(x) \cdot g(x) \tag{2.54}$$

The multiplication of the message $m(x)$ by x^{n-k} produces, as a result, a polynomial of degree $n-1$ where the first $n-k$ coefficients, now null, will then be occupied by parity bits.

From Eq. (2.54) it is clear that $m(x)x^{n-k}+r(x)$ is a multiple of the generator polynomial and it is therefore a codeword in accordance with Theorem 2.3.3.

Therefore the encoded word $c(x)$ is calculated as:

$$c(x) = m(x) \cdot x^{n-k} + r(x) \tag{2.55}$$

Suppose we have the BCH code *[15,7]*, already seen in the preceding section, generated by $g(x)=x^8+x^7+x^6+x^4+1$ and we have to encode the message $m=(1011011)$. The steps of the encoding are:

- $x^{n-k}\ m(x)=(000000001011011)$
- $r(x)=(x^{n-k}\ m(x))\ mod\ g(x)=(01101101)$
- $c(x)=(011011011011011)$

How these operations are effectively implemented in real applications is the topic of Chap. 8.

2.3.2.3 Decoding

The decoding operation follows three fundamental steps, as shown in Fig. 2.2.

- calculation of the syndromes;
- calculation of the coefficients of the error locator polynomial;
- calculation of the roots of the error locator polynomial.

During the transmission or the reading of the encoded message some errors may occur. Also the errors can be represented by a polynomial that has coefficient 1 in correspondence with every error's position:

$$E(x) = E_0 + E_1 x + \ldots + E_{n-1} x^{n-1} \tag{2.56}$$

Observe that, in order for the code to be corrector of t errors, at most t non-null coefficients are allowed in Eq. (2.56).

The read vector $R(x)$ is therefore:

$$R(x) = c(x) + E(x) \tag{2.57}$$

The first decoding step consists in calculating the $2t$ syndromes for the read message:

$$\frac{R(x)}{\psi_i(x)} = Q_i(x) + \frac{S_i(x)}{\psi_i(x)} \quad \text{with} \quad 1 \le i \le 2t \tag{2.58}$$

$$S_i(x) = Q_i(x) \cdot \psi_i(x) + R(x) \quad \text{with} \quad 1 \le i \le 2t \tag{2.59}$$

In accordance with Eq. (2.58) and Eq. (2.59), the received vector is divided by each minimal polynomial ψ_i forming the generator polynomial, thus getting a quotient $Q_i(x)$ and a remainder $S_i(x)$ called *syndrome*.

At this point the $2t$ syndromes must be evaluated into the elements $\beta, \beta^2, \beta^3, \ldots, \beta^{2t}$ whose ψ_i are the minimal polynomials. With reference to Eq. (2.60) this evaluation is reduced to an evaluation of the message received in $\beta, \beta^2, \beta^3, \ldots, \beta^{2t}$, since $\psi_i(\beta^i) = 0$ (for $1 \le i \le 2t$) by definition of minimal polynomial.

$$S_i(\beta^i) = S_i = Q_i(\beta^i) \cdot \psi_i(\beta^i) + R(\beta^i) = R(\beta^i) \tag{2.60}$$

Consequently, the i-th syndrome can be calculated either as a remainder of the division between the received message and the minimal polynomial ψ_i, then evaluated in β^i, or as the evaluation in β^i of the received message.

Observe that, in case no errors occur, the polynomial received is a codeword: therefore the remainder of the division of Eq. (2.58) is null and all the syndromes are identically null. On the other hand, verifying if the syndromes are identically null is a necessary and sufficient condition to understand if the read message is a codeword or if some errors occurred.

For binary codes we use the property:

$$S_{2i} = S_i^2 \tag{2.61}$$

It is therefore necessary to calculate only t syndromes. The syndromes are linked to the error positions, in fact:

$$S_i = R(\alpha^i) = c(\alpha^i) + E(\alpha^i) = E(\alpha^i) \tag{2.62}$$

By indicating the error positions with X and the number of errors that have really occurred with v (for a code corrector of t errors it must be $v \leq t$), we get Eq. (2.63).

$$\begin{aligned}
S_i &= E(\alpha^i) = x^{e_1} + x^{e_2} + \ldots + x^{e_v} \\
&= X_1^i + X_2^i + \ldots + X_v^i = \sum_{l=1}^{v} X_l^i
\end{aligned} \tag{2.63}$$

From Eq. (2.63) we get Eqs. (2.64) called *power-sum symmetric functions*:

$$\begin{aligned}
S_1 &= X_1 + X_2 + \ldots + X_v \\
S_3 &= X_1^3 + X_2^3 + \ldots + X_v^3 \\
&\vdots \\
S_{2t-1} &= X_1^{2t-1} + X_2^{2t-1} + \ldots + X_v^{2t-1}
\end{aligned} \tag{2.64}$$

Each method for solving these equations is a BCH decoding procedure. The direct method or *Peterson method* will be now described, since, although not used in applications, it is useful to clarify the theoretical steps of the decoding process.

Definition 2.3.14 It is defined as *error locator polynomial $\Lambda(x)$* the polynomial whose roots are the inverse of the error positions.

From the definition we have:

$$\Lambda(x) = \prod_{i=1}^{v}(1 - xX_i) \tag{2.65}$$

Observe that the degree of the error locator polynomial gives the number of errors that occurred. The degree of $\Lambda(x)$ is t at most, so, in the case more than t errors occur, the polynomial $\Lambda(x)$ could erroneously indicate t or less errors.
For each error X_j it must be:

$$1 + \Lambda_1 X_j^{-1} + \ldots + \Lambda_{v-1} X_j^{-v+1} + \Lambda_v X_j^{-v} = 0 \tag{2.66}$$

Multiplying by X_j^{i+v} we get

$$X_j^{i+v} + \Lambda_1 X_j^{i+v-1} + ... + \Lambda_{v-1} X_j^{i+1} + \Lambda_v X_j^i = 0 \tag{2.67}$$

then, by summing over j,

$$\sum_{j=1}^{v} X_j^{i+v} + \Lambda_1 \sum_{j=1}^{v} X_j^{i+v-1} + ... + \Lambda_v \sum_{j=1}^{v} X_j^i = 0 \tag{2.68}$$

$$S_{i+v} + \Lambda_1 S_{i+v-1} + ... + \Lambda_v S_i = 0 \tag{2.69}$$

which rewritten in matricial form becomes:

$$
\begin{pmatrix} S_{v+1} \\ S_{v+2} \\ S_{v+3} \\ \vdots \\ S_{2v-1} \end{pmatrix}
=
\begin{pmatrix}
S_1 & S_2 & S_3 & \cdots & S_v \\
S_2 & S_3 & S_4 & \cdots & S_{v+1} \\
S_3 & S_4 & \cdots & \cdots & S_{v+2} \\
\vdots & \vdots & \vdots & \vdots & \vdots \\
S_v & S_{v+1} & S_{v+2} & \cdots & S_{2v-1}
\end{pmatrix}
\cdot
\begin{pmatrix} \Lambda_v \\ \Lambda_{v-1} \\ \Lambda_{v-2} \\ \vdots \\ \Lambda_1 \end{pmatrix}
\tag{2.70}
$$

Consequently, once the syndromes are known, it is possible to find the coefficients of the error locator polynomial.

The last step of the decoding process consists in searching for the roots of the error locator polynomial. If the roots are separate and they are in the field, then it is enough to calculate their inverse to have the error positions. If they are not separate or they are not in the correct field, it means that the word received has a distance from a codeword greater than t. In this case an incorrigible error pattern occurred and the decoding process fails.

The general implementation for a BCH code will be presented in Chap. 8, while Chap. 9 will describe one of its particular application.

2.3.3 Reed-Solomon codes

2.3.3.1 Definitions

Reed-Solomon codes were described for the first time in 1960, in the article "Polynomial codes over certain finite fields" by Reed and Solomon. It is only later that the close link with BCH codes was understood.

There are a lot of ways to present Reed-Solomon codes; in the following we will prefer the one that more closely links them to BCH codes. In this way, we will be able to underline that Reed-Solomon codes are a subset of BCH codes q-ari.

Looking back to the Definition 2.3.12 of BCH codes we can affirm:

Definition 2.3.15 Reed-Solomon codes are BCH codes q^m-ari with $n=q^m-1$.

The first difference with BCH codes is the lack of the two fields of work $GF(q)$ and $GF(q^m)$, since they are coincident. This implies that the field where $g(x)$ has its coefficients and the one where $g(x)$ has its roots coincide.

This is the reason why binary Reed-Solomon codes do not exist. Their existence would imply $q=2$ and $m=1$, therefore an n equal to a single bit. Generally, in fact, Reed-Solomon codes correct symbols instead of bits.

The construction of the generator polynomial $g(x)$ results to be very simple, as the minimal polynomial in $GF(q^m)$ of an element β in $GF(q^m)$ is simply $x-\beta$.

Therefore, a generator polynomial of a Reed-Solomon code has the form

$$g(x) = \left(x - \beta^b\right) \cdot \left(x - \beta^{b+1}\right) \cdots \left(x - \beta^{b+2t-1}\right) \tag{2.71}$$

and the code has $k=q^m-1-2t$ information symbols.

Similarly to BCH codes, assuming $b=1$ we get narrow-sense codes, even though, in applications, the choice $b=0$ often simplifies the computational complexity.

Contrary to what said for BCH codes, where the minimum distance could be greater than the one designed, here the minimum distance is exactly $2t+1=n-k+1$. The importance of these codes lies exactly in this property which implies that, for a given correction capability, Reed-Solomon codes have the minimum number of redundancy symbols.

A particular case, and surely the most used in applications, concerns Reed-Solomon codes in $GF(2^m)$, where each symbol is composed by m bits.

Suppose, for example, we search for a code whose symbol is a byte, namely 8 bits. By definition of Reed-Solomon codes, the codeword length is fixed at $2^8-1=255$ bytes. In the particular case of a code able to correct 2 symbols, we need 4 parity bytes.

Now, suppose we wish to construct a Reed-Solomon code in $GF(8)$ corrector of 2 error symbols. The primitive polynomial of the field is x^3+x+1. The code is 7 symbols long, 3 of them are information symbols. The generator polynomial must have 4 roots in $GF(8)$:

$$\begin{aligned} g(x) &= (x - \alpha) \cdot \left(x - \alpha^2\right) \cdot \left(x - \alpha^3\right) \cdot \left(x - \alpha^4\right) \\ &= x^4 + \alpha^3 x^3 + x^2 + \alpha x + \alpha^3 \end{aligned} \tag{2.72}$$

The parity matrix of such code is:

$$H = \begin{pmatrix} 1 & \alpha & \alpha^2 & \alpha^3 & \alpha^4 & \alpha^5 & \alpha^6 \\ 1 & \alpha^2 & \alpha^4 & \alpha^6 & \alpha & \alpha^3 & \alpha^5 \\ 1 & \alpha^3 & \alpha^6 & \alpha^2 & \alpha^5 & \alpha & \alpha^4 \\ 1 & \alpha^4 & \alpha & \alpha^5 & \alpha^2 & \alpha^6 & \alpha^3 \end{pmatrix} \tag{2.73}$$

Using the primitive polynomial of the field it is possible to write the matrix of Eq. (2.73) in binary form, in a way analogous to what has been done with BCH codes in Eq. (2.52).

The code just described corrects 2 symbols, each composed by 3 bits. It is however wrong to deduce that such code corrects 6 bits, since, if errors occur in different symbols, the code is unable to correct them. Thus, the code is able to correct a number of bits varying from 2 to 6 depending on the position of the errors inside the symbols.

Due to the fact that they were born to correct symbols, Reed-Solomon codes are particularly suitable when burst type errors occur (Sect. 1.5.3).

We will now present the structure of encoding and decoding for Reed-Solomon codes, underlining the differences with binary BCH codes; because, being a generalization, the architecture is very similar.

The general structure of a Reed-Solomon code is that of Fig. 2.3.

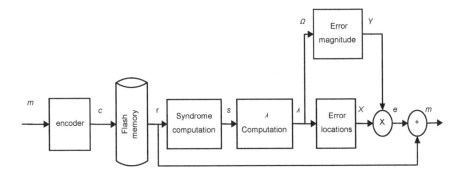

Fig. 2.3. Structure of a Reed-Solomon code

2.3.3.2 Encoding

The encoding and the decoding of Reed-Solomon code are described with reference to the particular case of codes in $GF(2^m)$. The encoding process is performed in a way completely analogous to the one described for BCH codes.

The only difference with the binary BCH case lies in the fact that the coefficients of the generator polynomial are not in $GF(2)$ anymore.

Suppose we have a Reed-Solomon code *[15,11]*. The code exists in $GF(2^4)$ and it is able to correct 2 symbols, each of 4 bits.

Using the decimal notation (see Appendix C), the generator polynomial of this code is:

$$g(x) = x^4 + 15x^3 + 3x^2 + x + 12 \tag{2.74}$$

Suppose we want to encode the message *m=(1 2 3 4 5 6 7 8 9 10 11)*. The steps of the encoding process are:

- $x^{n-k} m(x) = (0\ 0\ 0\ 0\ 1\ 2\ 3\ 4\ 5\ 6\ 7\ 8\ 9\ 10\ 11)$
- $r(x) = (x^{n-k} m(x)) \bmod g(x) = (3\ 3\ 12\ 12)$
- $c(x) = (3\ 3\ 12\ 12\ 1\ 2\ 3\ 4\ 5\ 6\ 7\ 8\ 9\ 10\ 11)$

The implementation details are postponed to Chap. 8.

2.3.3.3 Decoding

In accordance with Fig. 2.3, Reed-Solomon codes require a further step during the decoding process, compared to binary BCH codes, since it is not enough to know the error positions but it is also necessary to know their magnitude.

We will begin the description by generalizing what already said for binary BCH codes.

The evaluation of the syndromes is a procedure analogous to the one described for BCH codes, even if those are now function of both the error positions and the error magnitudes, as indicated in Eq. (2.75).

As for BCH codes, also for Reed-Solomon codes the syndromes can be computed as an evaluation of the received polynomial in α powers or as a remainder of a division between the received polynomial and the minimal polynomials, evaluated in α powers.

$$S_i = E(\alpha^i) = Y_1 X_1^i + Y_2 X_2^i + \ldots + Y_v X_v^i \tag{2.75}$$

where Y_i is the magnitude of the i-th error and X_i is the location of the i-th error.

We get a system of $2t$ equations in $2v$ unknowns:

$$
\begin{pmatrix} S_1 \\ S_2 \\ \vdots \\ \vdots \\ S_{2t} \end{pmatrix} =
\begin{pmatrix}
X_1^1 & X_2^1 & \cdots & X_v^1 \\
X_1^2 & X_2^2 & \cdots & X_v^2 \\
\vdots & \vdots & & \vdots \\
\vdots & \vdots & & \vdots \\
X_1^{2t} & X_2^{2t} & \cdots & X_v^{2t}
\end{pmatrix}
\cdot
\begin{pmatrix} Y_1 \\ Y_2 \\ \vdots \\ Y_v \end{pmatrix}
\tag{2.76}
$$

Note that, unlike the binary case, the equations are not power-sum symmetric equations. Besides, it is necessary to calculate $2t$ syndromes since the simplification of Eq. (2.61) refers only to the binary case.

Nevertheless, they can be solved in an analogous way, by defining the error locator polynomial $\Lambda(x)$. From Eq. (2.66) we get, multiplying by $Y_j X_j^{i+v}$:

$$Y_j X_j^{i+v} + \Lambda_1 Y_j X_j^{i+v-1} + \ldots + \Lambda_{v-1} Y_j X_j^{i+1} + \Lambda_v Y_j X_j^i = 0 \tag{2.77}$$

Then, summing over j:

$$\sum_{j=1}^{v} Y_j X_j^{i+v} + \Lambda_1 \sum_{j=1}^{v} Y_j X_j^{i+v-1} + \ldots + \Lambda_v \sum_{j=1}^{v} Y_j X_j^i = 0 \tag{2.78}$$

$$S_{i+v} + \Lambda_1 S_{i+v-1} + \dots + \Lambda_v S_i = 0 \tag{2.79}$$

which rewritten in matrix form becomes:

$$\begin{pmatrix} S_{v+1} \\ S_{v+2} \\ S_{v+3} \\ \vdots \\ S_{2v-1} \end{pmatrix} = \begin{pmatrix} S_1 & S_2 & S_3 & \cdots & S_v \\ S_2 & S_3 & S_4 & \cdots & S_{v+1} \\ S_3 & S_4 & \cdots & \cdots & S_{v+2} \\ \vdots & \vdots & \vdots & & \vdots \\ S_v & S_{v+1} & S_{v+2} & \cdots & S_{2v-1} \end{pmatrix} \cdot \begin{pmatrix} \Lambda_v \\ \Lambda_{v-1} \\ \Lambda_{v-2} \\ \vdots \\ \Lambda_1 \end{pmatrix} \tag{2.80}$$

With the same considerations made for BCH codes, once the syndromes are known, it is possible to calculate the error locator polynomial, whose roots are the inverse of the error positions.

At this point the decoding process is not finished, since it is necessary to know the magnitude of the errors. Once the v error locations are known, the system represented in Eq. (2.76) results to be a system of equations of $2v$ equations in v unknowns and it can be rewritten in the form:

$$\begin{pmatrix} S_1 \\ S_2 \\ \vdots \\ S_v \end{pmatrix} = \begin{pmatrix} X_1 & X_2 & \cdots & X_v \\ X_1^2 & X_2^2 & \cdots & X_v^2 \\ \vdots & \vdots & \ddots & \vdots \\ X_1^v & X_2^v & \cdots & X_v^v \end{pmatrix} \cdot \begin{pmatrix} Y_1 \\ Y_2 \\ \vdots \\ Y_v \end{pmatrix} \tag{2.81}$$

The decoding is complete solving for Y_i.

Also in this case the description of the implementation in real applications is postponed to Chap. 8.

Bibliography

E. R. Berlekamp, "Algebraic Coding Theory" in *New York: McGraw-Hill*, 1968.

R. E. Blahut, "Thory and Practice of Error Control Codes" in *Addison-Wesley*, 1984.

R. C. Bose, D. K. Ray-Chaudhuri, "On a Class of Error-Correcting Binary Group Codes", in *Information and Contribution*, Vol. 3, March 1960.

H. C. Chang, C. B. Shung, C. Y. Lee, "A Reed-Solomon product-code (RS-PC) decoder chip for DVD Applications" in *IEEE Journal of Solid State Circuits*, Vol. 36, February 2001.

R. T. Chien, B. D. Cunningham, J. B. Oldham, "Hybrid Methods for Finding Roots of a Polynomial with Application to BCH Decoding" in *IEEE Transactions on Information Theory*, March 1969.

C. K. P. Clarke, "Reed-Solomon Error Correction" in *British Broadcasting Corporation, R&D White Paper*, 2002.

G. Cohen, T. Mora, O. Moreno, "Applied Algebra, Algebraic Algorithms and Error-Correcting Codes: 10th International Symposium AAECC-10 San Juan de Puerto Rico", *Springer-Verlag*, 1993.

A. E. Heydtmann, J. M. Jensen, "On the Equivalence of the Belekamp-Massey and the Euclidean Algorithms for Decoding" in *IEEE Transactions on Information Theory*, Vol. 46, Nov. 2000.

A. Hocquenghem, "Codes correcteurs d'erreurs", *Chiffres*, Vol. 2, September 1959.

F. E. Hohn, "Applied Boolean Algebra: An Elementary Introduction", *The Macmillan Company*, 1966.

J. Hong, M. Vetterli, "Simple Algorithms for BCH Decoding" in *IEEE Transactions on Communications*, Vol. 43, August 1995.

J. N. Laneman, C. E. W. Sundberg, "Reed-Solomon Decoding Algorithms for Digital Audio Broadcasting in the AM Band" in *IEEE Transactions Broadcasting*, Vol. 47, June 2001.

S. Lin, D. J. Costello, "Error Control Coding: Fundamentals and Applications", *F. F. Kuo Editor*, 1983.

J. Massey, "Shift-Register Synthesis and BCH Decoding" in *IEEE Transactions on Information Theory*, Vol. 15, January 1969.

H. F. Mattson, T. Mora, T. R. N. Rao, "Applied Algebra, Algebraic Algorithms and Error-Correcting Codes: 9th International Symposium AAECC-9 New Orleans, USA", *Springer-Verlag*, 1992.

E. Mendelson, "Shaum's Outline of Theory and Problems of Boolean Algebra and Switching Circuits", *McGraw Hill*, 1970.

C. H. Papadimitriou, "Computational Complexity", *Addison-Wesley Publishing Company*, 1995.

W. W. Peterson, "Error-Correcting Codes" in *Cambridge, M.I.T. Press*, 1965.

I. S. Reed, G. Solomon, "Polynomial Codes over Certain Finite Fields", *Journal of SIAM*, Vol. 8, June 1960.

S. Sakata, "Applied Algebra, Algebraic Algorithms and Error-Correcting Codes: 8th International Conference AAECC-8 Tokio, Japan", *Springer-Verlag*, 1991.

H. Stichtenoth, "Algebraic Function Fields and Codes", *Springer-Verlag*, 1991.

B. Vucetic, Y. Jinhong, "Turbo Codes: Principles and Applications", *Kluwer Academic Publishers*, 2000.

S. B. Wicker, "Error Control for Digital Communication and Storage", *Prentice Hall*, 1995.

3 NOR Flash memories

3.1 Introduction

It was in 1965, just after the invention of the bipolar transistor by W. Shockley, W. Brattain and J. Bardeen, that Gordon Moore, co-founder of Intel, observed that the number of transistors per square centimeter in a microchip doubled every year. Moore thought that this tendency might be respected also in the years to come, and in fact the following years showed a doubling of the density of active components in an integrated circuit every 18 months. For example, in the 18 months that elapsed between the Pentium processor 1.3 and the Pentium-4, the number of transistors passed from 28 million to 55 million.

Today, a common table PC has processors with operating frequencies in the order of some gigahertz (1 GHz corresponds to about one billion operations per second) and a memory able to contain the information of several hundreds of gigaByte (GB).

In this scenario, a meaningful portion of the devices produced is represented by memories, one of the key components of all electronic systems.

Semiconductor memories can be divided into two major categories: RAM, acronym for Random Access Memories, and ROM, acronym for Read Only Memories: RAM loses its content when the power supply is switched off, while ROM virtually holds it forever. The category of non volatile memories includes all the memories whose content can be changed electrically but it is held when the power supply is switched off. These are more flexible than the original ROM, whose content is defined during the manufacturing and cannot be changed by the consumer anymore.

The history of non-volatile memories begins in the seventies, with the introduction of the first EPROM memory (Erasable Programmable Read Only Memory). Since then, non-volatile memories have always been considered one of the most important families of semiconductors and, up to the nineties, their interest was tied up more to their role as a product of development for new technologies than to their economic value. Since the early nineties, with the introduction of non-volatile Flash memories into portable products like mobile phones, palmtop, camcorders, digital cameras and so on, the market of these memories has had a staggering increase.

Over the years, the evolution of technology for non-volatile memories called floating gate memories has allowed a reduction in the area of a single memory cell, which, from 1993 up to today, has passed from 4 to 0.4 square micrometers[1]. The

---

[1] T. Tanaka, "NAND Flash Design", International Solid-State Circuits Conference, Memory Forum, Feb. 2007.

storage capacity of memory devices has passed from 16 Mb in 1993 to 16 Gb in 2008, with a variation equal to 1000 times, in perfect accordance with Moore's law.

Let's now try to better understand the Flash memory based on a floating gate cell whose section is shown in Fig. 3.1: the potentials used for hot electron program are shown. We make a MOS transistor with two overlapping gates rather than only one: the first completely isolated in the oxide, the second, instead, contacted to form the gate terminal. The isolated gate, completely surrounded by the oxide, constitutes an excellent "trap" for the electrons placed on it, such as to guarantee charge retention for some tens of years. The operations which allow moving the electrons into the isolated gate and then removing them, when necessary, are called program and erase. These operations allow us to modify a macroscopic electric parameter which corresponds to the threshold voltage of the memory cell, that is the value of the voltage to be applied to the control gate so that the memory cell can be considered switched on, and which is used to discriminate the logic value to be stored. Figure 3.2 contains the schematic representation of a single cell and its relative layout.

The memory cells are packed to form a matrix so as to optimize the space on silicon.

In Fig. 3.3 there is a typical disposition for a Flash process with NOR-type architecture. Figure 3.3 shows the electric schematic diagram of a group of cells together with the layout that will be used for the real physical realization of the cells.

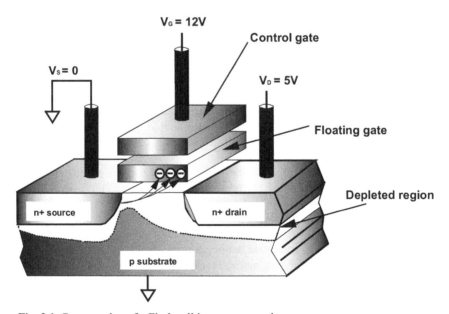

Fig. 3.1. Cross-section of a Flash cell in program mode

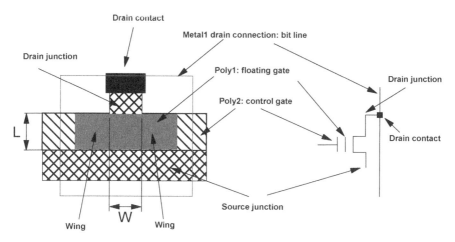

Fig. 3.2. Layout of the non-volatile cell and corresponding schematic

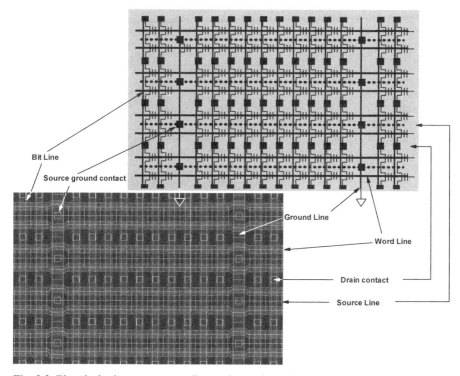

Fig. 3.3. Electrical scheme corresponding to the portion of NOR sector

The connections denominated wordlines connect the gates of the memory cells, while those called bitlines connect the drains. The drain contact, in Fig. 3.3 it is the black little square, is used to connect the silicon area underlying and it is shared by two cells. With the type of disposition adopted we can reach densities of millions of cells in few square millimeters.

Flash memories are non-volatile memories characterized by the fact that the erase operation (writing of the logic "1") must occur simultaneously on a set of cells, called sector or block; the program operation instead (writing of the logic "0") acts on the single cell. The fact that the erase operation can take place only on an entire sector allows the matrix to be designed in a more compact form. For the Flash memories currently on the market, the matrix architectures by far the most used are called NOR and NAND. The common element between the two architectures is the non-volatile memory cell. Table 3.1 shows the most important parameters of the two types of memory.

Table 3.1. NOR and NAND performances

	NOR 1b/cell	NOR 2bits/cell	NAND 1b/cell	NAND 2bits/cell
Page size	32 B	1 KB	4 KB	4 KB
Read access time	<70 ns	<80 ns	25 µs	50 µs
Program time	9 µs/Word	4.2 µs/Word	150 µs/page	600 µs/page
Erase time	1 s/sector	1 s/sector	2 ms/sector	2 ms/sector

We define as output parallelism the number of bits that the memory supplies to the output simultaneously. The page size is the number of bits accessed during read/write inside the memory.

The purpose of the present chapter is to explain the basic functionalities of a NOR Flash memory; the NAND Flash functionalities are described in Chap. 4.

3.2 Read

One of the most important specifications for any memory, not only of Flash type, is certainly the data access time.

For access time we mean the time interval between any commutation of the addresses and the instant in which the data addressed is made available to the output. Such time is commonly defined with the term "asynchronous access time". The term "asynchronous" is used to distinguish this read mode from another, said synchronous, characterized by the fact that the data must be delivered to the output in a way synchronous to an external clock.

The reading of a non-volatile memory cell is performed by applying specific voltages to the terminals of the cell itself, and by sensing the current that flows in it. NOR and NAND clearly differ in the way they sense such current. Let's first analyze the NOR case; for NAND refer to Chap. 4.

In NOR-type Flash memories, the reading of the matrix cell takes place in a differential way, by comparing its current to the one of a reference cell physically identical to the cells of the matrix and biased with the same V_{GS} and V_{DS}.

In the case of 1bit/cell memories, the I_{DS}-V_{GS} electric characteristics of the written cell (logic "0") and of the erased one (logic "1") result separate, as illustrated in Fig. 3.4; this is because the two cells have different V_{THS} and V_{THC} threshold voltages.

With equal V_{WL} read voltage applied to the control gate, the cell "written" (logic "0") conducts an I_{DSS} current smaller than the I_{DSC} "erased" one (logic "1"). The reference characteristic must be placed between these two. In this way, if we apply a V_{GS} common to all (V_{WL} in Fig. 3.4), the cell will be recognized as erased if its current results higher than the I_{DSR} reference one, otherwise as written. As a matter of fact, the current of the cells is converted into voltage through a current voltage converter (I/V), i.e. a circuit that provides an output voltage which depends on the input value of the current. These voltage values are then compared by a voltage comparator, at whose output we have the logic levels "0" and "1" (see Fig. 3.5), corresponding to the cell state.

It is also important, during the read operation, to prevent the drain of the cells from being at a too elevated potential in order to avoid spurious programming. It is usually better to limit the drain potential at 1 V. On top of the I/V converter it is therefore necessary to have a block that during the read operation does not allow the drain of the cells to exceed such value.

In the majority of the cases, the circuitry in charge of the read operation (commonly named *sense amplifier*) cannot disregard the following fundamental blocks:

- I/V converter;
- drain voltage limiter (typically at 1 V);
- output comparator.

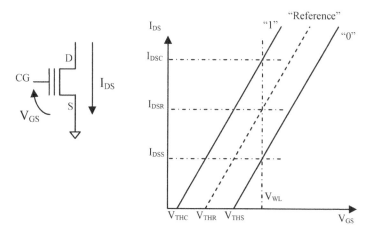

Fig. 3.4. Current/voltage characteristics of a Flash cell as the threshold voltage changes

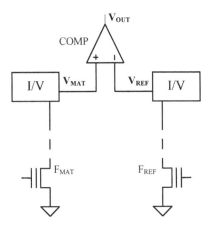

Fig. 3.5. Block diagram of a circuit used to compare two currents

One of the simplest circuit implementations of a sense amplifier is shown in Fig. 3.6: on the left hand side we have the matrix cell (F_{MAT}), on the other side we have the reference cell (F_{REF}). They are both biased with the same gate voltage (V_{WL}) and they are accessed through the column decoder, made up by NMOS transistors (M2, M3, M5 and M6). In Fig. 3.6 the two NOR logic gates together with the transistors M1 and M4 act as voltage limiters for the drains of the memory cells. The SAEN# signal enables the sense amplifier. When the circuit is switched on (SAEN# low), the drain terminals (i.e. bitline) of the memory cells node are at ground and the output of the NOR gate equals VDD. Thus, M1 has V_{GS} that equals VDD, which is the maximum voltage available to the circuit. The charging of the bitline can therefore take place at the maximum current and in the minimum time. The switching threshold voltage of the NOR gates is designed in order to prevent the bitlines from overcoming the limit of 1 V. The NOR gate in the feedback loop operates around its switching threshold voltage: therefore, a DC (Direct Current) consumption is present. For this reason, an enable signal, SAEN#, is introduced to keep the sense amplifiers "on" for the shortest time needed.

The I/V converter is made with a resistive load. The output voltages of the I/V converter, indicated with V_{MAT} and V_{REF}, are compared by the COMP voltage comparator. More complex circuit structures and new architectures have come out during the years in order to improve the performances of the sense amplifier and to reduce the read time more and more.

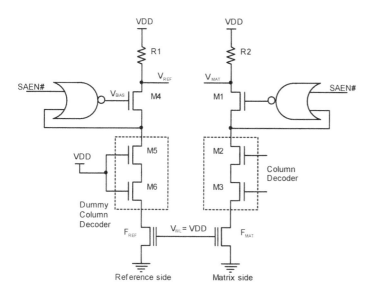

Fig. 3.6. Fundamental diagram of a *sense amplifier*

A considerable improvement is achieved with the technique called "equalization": before the reading operation takes place, the V_{MAT} and V_{REF} voltages are "equalized" (that is, forced) to the same value. In this way, we prevent excessive excursions of the voltages, in case, at the beginning of the read operation, such voltages are by far different from their final value. Reducing the voltage swing means, at the end, faster operations.

Another method of equalization is carried out on the drain nodes of the matrix cell and of the reference cell. This type of equalization, apparently simple in circuit terms, requires the use of a number of reference cells equal to the number of matrix cells read in parallel. For this type of equalization to be really effective, it is important for the reference cell to be under the same drain load conditions of the matrix cell. For this reason we emulate the capacitive load of the matrix cell on the branch of the reference cell; a real bitline is usually used to obtain the complete matching.

With this type of structures it is possible to carry out dynamic differential readings, which have the advantage to obtain the correct voltage separation of the V_{MAT} and V_{REF} voltages as soon as the equalization phase ends: that allows, at equal sensitivity of the COMP comparator, a much faster reading than in the classic case, where we need to wait for the V_{MAT} and V_{REF} voltages to stabilize.

Figure 3.7 clearly shows the benefit of the dynamic differential reading. Typically, the reading phase for the comparison of the current of the two cells takes approximately $10 \div 20$ ns.

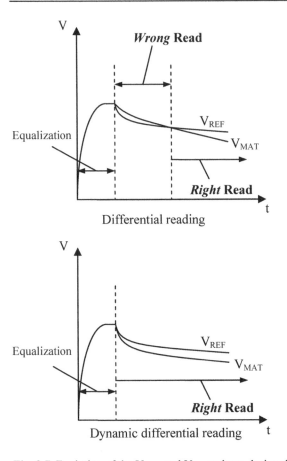

Fig. 3.7. Evolution of the V_{MAT} and V_{REF} voltage during the read operation

3.3 Program

Writing the information in the memory cell, i.e. programming, refers to an operation through which the electrons are brought from the substrate to the floating gate; in this way the threshold voltage of the cell increases.

In the introduction we saw how the number of bits programmed per second for a NAND memory is definitely greater if compared to that of a NOR memory. The reason for this difference is mainly linked to the fact that NAND and NOR use two different physical phenomena.

In this section we describe how the program operation for NOR memories takes place.

The writing of information in NOR-type cells is performed through the *Channel Hot Electron* mechanism.

Applying 5÷6 V between the drain and source regions of the memory cell, the resulting longitudinal electric field produces electrons having an energy greater than the one they would have inside the crystalline lattice at thermal equilibrium. These high energy electrons are called *hot electrons*. The most considerable part of the phenomenon of hot electrons generation occurs where the electric field is maximum, which is in the depletion region in proximity of the drain.

Some hot electrons succeed in acquiring enough energy (higher than 3.1 eV) to overcome the potential barrier at the tunnel oxide interface.

At this point, the application of a high voltage to the control gate has the purpose to create a transversal electric field to favor the injection of the electrons from the channel to the oxide and their gathering in the floating gate.

The process of hot electron injection does not continue endlessly, but it gradually decreases as the negative charge accumulates into the floating gate. In fact, such charge tends to reduce the gate potential and therefore to reduce its attraction property progressively. During the collisions, not only electrons are generated, but also holes, and while the first ones are collected at the drain giving origin to drain current (superior to the source one), the holes are collected in the substrate.

The program operation via hot electrons is a fast operation, but the required current is definitely high. Obviously, the greater the program parallelism we want to guarantee is (number of cells we want to program at the same time) the greater the power consumed is. For example, in the hypothesis of programming 64 cells, we need to produce 3.2 mA inside the chip, assuming each cell absorbs 50 uA. Thus, programming an elevated number of cells with this mechanism may become a heavy operation especially in terms of area required to build the peripheral circuits that allow the generation of such elevated currents.

The voltages that are typically applied to the cell during the writing operation are:

- 4.5 V on the drain of the cell;
- 9 V on the gate;
- 0 V on the source and body.

The choice of the values for these voltages depends on the technology used. The cell must, in fact, have a good programming speed and a good strength against parasitic effects such as the *drain turn on*, the *snap back*, the *soft programming* during read and the *soft erasing* during erase. It is therefore mandatory to consider aspects such as the channel length, the programming efficiency of the cell, the drain current and the process variations. The precision in the application of the voltages is also fundamental: variations in their values during the operation may cause a variation in the drain current and therefore a variation in the ability of the cell to accumulate electrons in a fixed time.

After a modify operation of the content of the floating gate (both write and erase) we get some threshold voltage distributions. This is due to different factors such as geometric dissymmetry, process variations, supply voltage variations, source and drain modulations. Nevertheless, it is important for the distributions to have a well precise and controlled width in order to be correctly allocated in the operating window of the memory cell. This requirement has naturally been exasperated by the introduction of multi-level memories, that is, memories in which more bits are stored in the same cell. In fact, as the number of bits increases, the number of distributions to be positioned in the operating window increases too. If, for example, the bits stored in the cell are 2, then we have 4 distributions to allocate (Fig. 3.8). Figure 3.9 shows how the Fig. 3.4 is then modified for 2bits/cell memories.

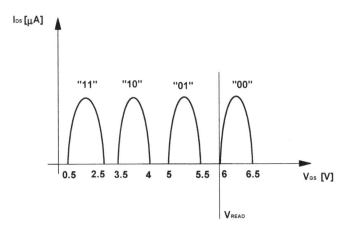

Fig. 3.8. Position of the threshold voltages for a cell containing 2 bits

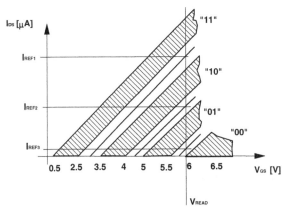

Fig. 3.9. Read operation requires 3 reference cells (currents)

The width of the distributions can actually be contained by choosing a suitable program algorithm: for example, a *Program & Verify* algorithm where the program pulse is followed by a verify operation. During this last phase we check that the programmed cells have reached the required threshold voltage. In the case of NOR memories we compare the content of the matrix cell with that of a reference cell: it is a question of a reading. As a matter of fact, two factors differentiate the read from the verify operation: the reference cell used and the time required.

In practice, we try to program the cells in such a way that they are all above the reference used in read. In order to increase its precision, the verify operation is usually longer than a normal read. After the verify operation, the programmed cells (in other words, cells which have reached the required threshold voltage) are left in that position, while another program pulse is applied to the remaining ones. The algorithm ends either when all the cells are programmed (in this case the program operation had a successful conclusion) or when the attempts available to program the cells have been used up (in this case the algorithm did not have a successful conclusion and this information is given to the external world).

To prevent a cell from being programmed it is therefore enough to make sure that it does not have high voltages on the drain and on the gate simultaneously. For the way the cells are connected in the memory matrix, however, we have cells where a high voltage is applied to the gate and cells where a high voltage is applied to the drain (Fig. 3.10).

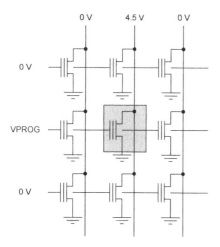

Fig. 3.10. Biasing of NOR matrix during a program operation. The cell under program is highlighted

Just because of these applied potentials, the cells can suffer from different disturbs. The cells which share the same bitline with the programmed one suffer from the so-called *drain disturb*: because of the potential seen on the drain, programmed cells (therefore with negative charge accumulated in the floating gate) may manifest a tendency to loose the accumulated charge. Instead, the cells which share the same wordline with the programmed one have their gates at a high potential, while their drains are grounded. The effect of this biasing on the erased cells is that of a program operation by means of Fowler – Nordheim tunneling; in other words, we may have an increase in their threshold voltages. On the contrary, on programmed cells, the effect of having a high voltage at the gate is that of an electron injection from the floating gate to the control gate and a consequent decrease of the threshold voltage of the cell.

3.4 Erase

The physical mechanism used for the erase operation in Flash memories, both NOR and NAND, is the Fowler-Nordheim tunneling.

To trigger the erase operation by means of Fowler-Nordheim tunneling it is necessary to apply an elevated voltage difference on the tunnel oxide. In first generation NOR memories this occurred by applying a high voltage to the source terminal (18 V represented a typical value), setting the control gate grounded and leaving the drain electrode floating. The body of the cell is grounded because this terminal coincides with the substrate of the whole device. In this way the Fowler-Nordheim tunneling occurs between gate and source, by extracting electrons from the floating gate: this is equivalent to making the threshold voltage of the cell more negative. With this kind of erase, we work very close to the source/substrate junction breakdown (as a matter of fact we are exactly in breakdown at the beginning of the erasure).

To solve this problem, in the second generation, the high voltage was split between the gate and the source terminals: gate electrode was negative in order to reduce the voltage on the source junction. However, this poses the important problem of producing on-chip voltages of negative value.

In any case, in the biasing schemes just described, the phenomenon known as *band-to-band tunneling* increases the leakage current through the source/substrate junction: so that a non negligible current consumption is associated with the tunneling. Besides, the band-to-band tunneling degrades the cell reliability because of the injection of holes into the floating gate.

In third generation memories, the most recent, the problem is solved by placing the matrix in triple well (Fig. 3.11): with such a structure, it is possible to bias the substrate, that is the isolated p-well (ip-well), with a high voltage. In such a way the electrons are extracted along the whole channel without further parasitic contribution from the source junction. As a result, the current consumption is reduced of about three orders of magnitude.

From Table 3.1 we can see that the erase times of a block differ of three orders of magnitude between NAND and NOR architectures. In fact, even thought the physical mechanism adopted is the same, the architectural differences, due to the different technical specifications to be satisfied, have a heavy impact on the type of erase algorithm which has to be implemented. Such differences are determined by how much the width and the position of the distribution of the erased cells need to be checked.

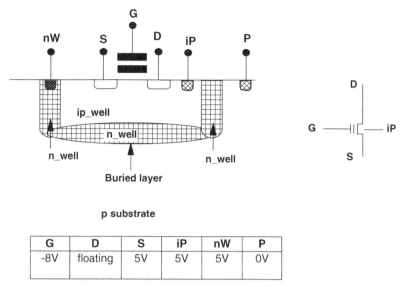

Fig. 3.11. Memory array in triple well

G	D	S	iP	nW	P
-8V	floating	5V	5V	5V	0V

In recent NOR Flash memories, each sector is built in its own triple well. The erase algorithm has to manage the voltages and their transitions to the terminals of the cells – addressed and not – in a correct way, and, as a matter of fact, it is much more complex than the simple application of a voltage pulse that triggers the tunneling.

At the end of the electric erase, the threshold voltage distribution of the cells can spread in the range of negative values: in the NOR architecture this is not acceptable. Obviously, the less erased cells must be sufficiently low to ensure a suitable distance between the programmed level and the erased one; on the contrary, the most erased cells cannot be moved downward indiscriminately. By construction, in fact, in the NOR architecture a number of cells insist on the same bitline. The un-addressed wordlines are grounded: this is indeed a switch off condition only if the cells have threshold voltages higher than zero. On the contrary, if the unselected cells are not completely switched off, a sub-threshold current flows along the bitline. Even if the contribution of the sub-threshold current of a single cell is relatively small, it can be multiplied by the number of cells insisting on the same bitline (128, 256 or also 512). The circuit described in Fig. 3.6 is not able to distinguish between the current of the selected cell and the leakage current of the bitline. As a result, a reduction of the read margins takes place.

To avoid the leakage contribution, it would be necessary to bias the unselected wordlines with a negative voltage during read: this is not done since it costs in terms of circuital complexity and of time spent to access the memory. In order to keep the typical read speed of the NOR architecture, we have to eliminate the leakage problem studying a specific algorithm.

To avoid making (all or partially) the threshold voltage distribution negative with a single erase pulse, the erase operation is executed in different phases. One of these phases is the *electrical erase* which is performed applying a voltage V_B on the body (ip-well) of the selected sector, keeping the gates at ground. Also in this case, a series of pulses is used together with verify operations in order to precisely control the distribution positioning. At each erase pulse, the voltage V_B is increased compared to the previous one. The following equation links the threshold voltage shift ΔV_{TH} after the erase pulse with the voltage difference applied on the last 2 electrical erase pulses (*n* and *n–1*):

$$\left| \Delta V_{TH} \right| = V_B(n) - V_B(n-1) \tag{3.1}$$

The erase verify consists in checking that the cell current is enough to well distinguish the erased cells from the programmed ones. For the purpose we use the comparison with a reference cell, called *erase verify* (EV). The sequence of erase pulses followed by erase verifies continues until all the cells of the sector have at least the same current of the EV reference. The erase pulse is applied to the whole sector in a non-selective way but not all the cells are erased at the same speed. When the slowest cells have eventually reached the erase verify threshold, the fastest ones, instead, will be heavily negative, i.e. *depleted*. To recover the cells in this condition, the algorithm enters a second phase, called *soft program*: using a *depletion verify*, DV, the depleted cells are reprogrammed in order to suppress the leakage on the bitline. The soft program routine applies program pulses (channel hot electrons) until all the cells of the sector have exceeded the threshold of the DV reference cell. If negative wordline voltages are available, the depletion verify can be done by applying negative voltages to the unselected wordlines, thus cutting down the cell leakage during verify.

All these reference cells are placed in a special reference matrix and their threshold voltages are appropriately set during the factory tests. Figure 3.12 summarizes what we said about the reference cells and their position relatively to the distributions for the erased and programmed matrix cells.

Coming back to the setting of the EV reference, we cannot choose a too low threshold value, assuming an advantage in terms of read margins: having a lower EV means a greater erase of the whole sector and the recovering of a greater number of depleted cells. A situation of this kind certainly damages both the soft program time (which, as we will see, represents the preponderant part of the algorithm) and the power consumption at the beginning of the soft program itself.

When the customer decides to erase a sector, the state of the cells inside can be the most varied: it is possible, for example, that few cells have been programmed, and that a lot are still erased. A cycling under these conditions is a cause of stress for the cells already erased; besides, a too wide initial distribution gives origin to a larger erase distribution and with more leakage. For this reason the real electrical erase is preceded by a non-selective program of the entire sector to a voltage value that is considered sufficient to contain the erase magnitude. This operation is called preconditioning or *program all 0*.

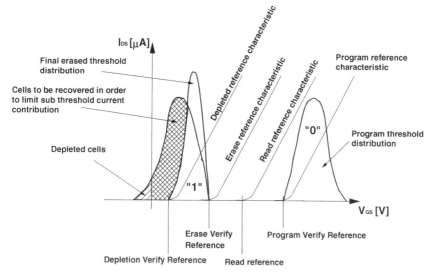

Fig. 3.12. Summary of the cells inside the Reference Matrix and their relative position to the "1" and "0" distributions

The erase sequence of a NOR architecture is shown in Fig. 3.13, together with the effects that the various phases have on the distribution.

As we have seen, the erase algorithm in NOR Flash memories is very complex; furthermore all these phases must be linked in the correct way, especially because we have to manage big capacitive loads charged with elevated potentials. For example, to allow the execution of the erase verify after an erase pulse, the ip-well and the source of the sector must be brought back from V_B to ground, while the corresponding n-well must be discharged from the high voltage to VDD.

The discharge of these nodes must be executed with special care. During each erase pulse the gates of the cells are charged to about –8 V, the ip-well is charged to about 10 V and the source and drain terminals, being floating, are charged to a high voltage (the exact value depends on the ip-well voltage and on the capacitive loads involved). If the gate is brought back to ground starting from these conditions and without care, the gate-drain parasitic capacitance "pushes" the drain node even higher (of a value equal to the proportion between the capacitances associated to the gate and drain nodes), thus risking a breakdown. To avoid this problem, when passing from erase pulse to verify, the ip-well is first discharged to ground; then, the drains are brought back to ground by activating the column decoding; afterwards, the source junction and the n-well are discharged to ground and to VDD respectively. Finally, the gates are brought back to ground, and will then be charged to the erase verify voltage. All these operations involve elevated parasitic capacitance and cannot be too fast: both to prevent a snap-back of the transistors responsible for the discharge and to avoid the noise on the ground level (*ground bounce*) which could determine the spurious commutation of the circuits.

Let's now look at the erase performance. As an example, we take a NOR Flash with 1 Mbit sectors and 128-bits pages during program. Assuming a typical erase time of 1 s per sector; we can have the following breakdown:

- the program all 0 part takes a negligible time, of the order of some hundreds of µs (they are just a few program pulses and moreover without verify);
- about 250 ms are devoted to the erase/erase verify sequence;
- about 750 ms, the most considerable time, is spent for the soft program and the depletion verify.

All these stratagems are necessary: the voltages and the times in which they are applied must be selected with the maximum care, because neither at time zero neither during the life of the device (100 Kcycles of program/erase admitted and a 10 year-old life) erase failures are tolerated.

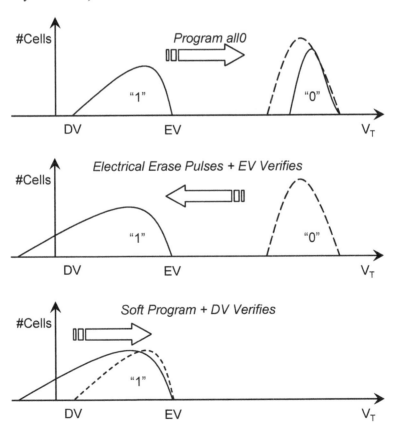

Fig. 3.13. The phases of a NOR erase algorithm and their effects on the erased distribution

Bibliography

B. K. Ahuja, "An Improved Frequency Compensation Technique for CMOS Operational Amplifiers", *Journal of Solid-State Circuits*, Vol. SC-18, pp. 629–633, November 1983.

L. A. Akerst, "An Analytical Expression for the Threshold Voltage of a Small Geometry MOSFET", *Solid State Electronics*, Vol. 24, pp. 621–627, 1981.

M. Annaratone, "Digital CMOS Circuit Design", *Kluwer Academic Publishers*.

H. Arakawa, "Address Decoder Circuit for Non-Volatile Memory", *United States Patents* 5,039,882, August 13, 1991.

S. Aritome, "Advance Flash Memory Technology and Trends for File Storage Application", in *IEDM Tech. Dig.*, pp. 763–766, 2000.

S. Aritome, R. Shirota, G. Hemink, T. Endoh, F. Masoka, "Reliability Issues of Flash Memory Cells", *Proc. IEEE*, Vol. 81, No. 5, pp. 776–788, May 1993.

S. Atsumi et al., "A 3.3V-only 16 Mb Flash Memory with Row-Decoding scheme", 1996 *IEEE Int. Solid-State Circuits Conf. Dig. Tech. Papers*, pp. 42–43, February 1996.

S. Atsumi et al., "A Channel-Erasing 1.8 V-Only 32-Mb NOR Flash EEPROM with a Bitline Direct Sensing Scheme", *IEEE Journal of Solid-State Circuits*, Vol. SC-35, pp. 1648–1654, November 2000.

C. Auricchio et al., "A Triple-Well Architecture for Low Voltage Operation in Submicron CMOS Devices", (Eds.), *Proc. ESSDERC 96*, Bologna, Italy, p. 613, 1996.

M. Bauer et al., "A multilevel-cell 32 Mb Flash Memory", in *1995 IEEE Int. Solid-State Circuits Conf. Dig. Tech. Pap.*, pp. 132–133, February 1995.

H.P. Belgal et al., "A New Reliability Model for Post-cycling Charge Retention of Flash Memories", *Proc. IRPS*, pp. 7–20, 2002.

A. Bellaouar et al., "Bootstrapped Full-Swing BiCMOS/BiNMOS Logic Circuits for 1.2–3.3 V Supply Voltage Regime", *IEEE Journal of Solid-State Circuits*, Vol. 30, No. 6, June 1995.

R. Bez, "Introduction to Flash Memory", *IEEE Proceeding of the*, Vol. 91, No. 4, pp. 489–502, April 2003.

R. Bez et al., "Depletion Mechanism of Flash Cell Induced by Parasitic Drain Stress Contidion", VLSI Technology Symposium, 1994.

R. Bez et al., "Introduction to Flash Memory", *Proceedings of the IEEE*, Vol. 91, pp. 554–568, 2003.

C. S. Bill el al., High Voltage Charge Pumps with Series Capacitances, *USA Patent* No. 5,059,815.

A. Brand et al., "Novel Read Disturb Failure Mechanism Induced by FLASH Cycling", *IEEE IRPS*, 1993.

W. D. Brown, J. E. Brewer, eds., "Nonvolatile Semiconductor Memory Technology", *New York, NY: IEEE Press*, 1998.

G. Campardo, R. Micheloni, "Row Decoder for a Flash-EEPROM Memory Device with the Possibility of Selective Erasing of a Sub-Group of Rows of a Sector", *USA Patent* No. 6,122,200.

G. Campardo, R. Micheloni, "Architecture of Non Volatile Memory with Multi-bit Cells", 12th Bi-annual Conference June 20–23, INFOS2001, Invited paper, 2001.

G. Campardo, R. Micheloni, "Architecture of Non volatile Memory with Multi-Bit Cells", *Elsevier Science, Microelectronic Engineering*, Vol. 59, No. 1-4, pp. 173–181, November 2001.

G. Campardo, R. Micheloni, "Scanning the Special Issue on Flash Memory Technology", *IEEE Proceeding of the*, Vol. 91, No. 4, pp. 483–488, April 2003.

G. Campardo, R. Micheloni, S. Commodaro, "Line Decoder for Memory Devices", *USA Patent* No. 6,018,255.

G. Campardo, R. Micheloni, S. Commodaro, "Low Supply Voltage Nonvolatile Memory Device with Voltage Boosting", *USA Patent* No. 5,903,498.

G. Campardo, R. Micheloni, S. Commodaro, "Method and Circuit for Reading Low-Supply-Voltage Nonvolatile Memory Cells", *USA Patent* No. 6,128,225.

G. Campardo, R. Micheloni, M. Maccarrone, "Circuit and Method for Generating a Read Reference Signal for Nonvolatile Memory Cells", *USA Patent* No. 5,805,500.

G. Campardo, R. Micheloni, M. Maccarrone, "Read Circuit and Method for Nonvolatile Memory Cells with an Equalizing Structure", *USA Patent* No. 5,886,925.

G. Campardo, R. Micheloni, D. Novosel "VLSI-Design of Non-volatile Memories", Springer Series in ADVANCED MICROELECTRONICS, 2005.

G. Campardo et al., "Method and Circuit for Dynamic Reading of a Memory Cell at Low Supply Voltage and with Low Output Dynamics", *USA Patent* No. 6,639,833.

G. Campardo et al., "Method and Circuit for Generating an ATD Signal to Regulate the Access to a Non-volatile Memory", *USA Patent* No. 6,075,750.

G. Campardo et al., "A 40 mm2 3V 50 MHz 64 Mb 4-level Cell NOR Type Flash Memory", *2000 ISSCC, San Francisco*.

G. Campardo et al., "Circuit Structure for Providing a Hierarchical Decoding in Semiconductor Memory Devices", *USA Patent* No. 6,515,911.

G. Campardo et al., "40- 3-V-only 50-MHz 64-Mb 2-b/cell CHE NOR Flash Memory", *IEEE Journal of Solid-State Circuits*, Vol. SC-35, No. 11, pp. 1655–1667, November. 2000.

G. Campardo et al., "An overview of Flash Architectural Developments", *IEEE Proceeding of the*, Vol. 91, No. 4, pp. 523–536, April 2003.

P. Cappelletti, A. Modelli, "Flash Memory Reliability", in *Flash Memory*, P. Cappelletti et al., Ed Kluwer, 1999.

P. Cappelletti et al., Eds., "*Flash Memories*", Norwell, MA: Kluwer, 1999.

E. Charbon, R. Gharpurey, P. Miliozzi, R. G. Meyer, A. Sangiovanni-Vincentelli, "SUBSTRATE NOISE – Analysis and Optimization for IC design", *Kluwer Academic Publishers*, 2001.

R. L. M. Dang, N. Shigyo, "Coupling Capacitance for Two-Dimensional Wires", *IEEE Electron Deviced Letters*, Vol. EDL-2, No. 8, pp. 196–197, August 1981.

S. D'arrigo et al., "A 5V-Only 256K Bit CMOS Flash EEPROM", *ISSCC 89*, pp. 132–133.

S. H. Dhong et al., "High Speed Sensing Scheme for CMOS DRAM's", *IEEE Journal of Solid-State Circuits*, Vol. 23, No. 1, pp. 34–40, February 1988.

G. Di Cataldo, G. Palumbo, "Double and Triple Charge Pump for Power IC Dynamic Models Which Take Parasitic Effects into Account", *IEEE Transaction Circuits System*, Vol. CAS-40, pp. 92–101, February 1993.

J. Dickson, "On-Chip High Voltage Generation in MNOS Integrated Circuits Using an Improved Voltage Multiplier Technique", *IEEE Journal of Solid-State Circuits*, Vol. SC-11, No. 3, pp. 374–378, June 1976.

C. Dunn et al., "Flash EEPROM Disturb Mechanisms", in *Proc. Int. Rel. Phys. Symp.*, pp. 299–308, April 1994.

B. Eitan and D. Frohman-Bentchkowski, "Hot Electron Injection into the Oxide in n-channel MOS Devices", *IEEE Trans. Electron Devices*, Vol. ED-28, pp. 328–340, March 1981.

D. Elmhurst et al., "A 1.8 V 128 Mb 125 MHz Multi-level Cell Flash Memory with Flexible Read While Write", ISSCC Dig. Tech. Papers, pp. 286–287, February 2003.

D. Frohman-Bentchkowsi, "Memory Behavior in a Floating Gate Avalanche-Injection MOS (FAMOS) Structure", *Appl. Phys. Lett.*, Vol. 18, pp. 332–334, 1971.

D. Frohman-Bentchkowsi, "FAMOS-A New Semiconductor Charge Storage Device", Solid State Electron, Vol. 17, pp. 517–520, 1974.

G. Ginami. et al., "Survey on Flash Technology with Specific Attention to the Critical Process Parameters Related to Manufacturing", *IEEE Proceeding of the,* Vol. 91, No. 4, pp. 503–522, April 2003.

P. R. Gray, R.G. Meyer, "MOS Operational Amplifier Design- A Tutorial Overview", *IEEE Journal of Solid State Circuits*, Vol. SC-17, No. 6, pp. 969–982, December 1982.

R. Gregorian, G. C. Temes, Analog Mos Integrated Circuits for Signal Processing, *J. Wiley & Sons*, 1986.

M. Grossi et al., "Program Schemes for Multilevel Flash Memories", *IEEE Proceeding of the*, Vol. 91, No. 4, pp. 594–601, April 2003.

S. S. Haddadi et al., Flash E2PROM Array with Negative Gate Voltage Erase Operation, *U.S. Patent* No. 5,077,691, October 23, 1989.

T. P. Haraszti, CMOS Memory Circuits. *Boston, MA: Kluwer Academic Publishers*, Ch. 5, 2000.

H. Hidaka et al., "Twisted Bit-Line Architectures for Multi-Megabit DRAM's", *IEEE Journal of Solid-State Circuits*, Vol. 24, No. 1, pp. 21–27, February 1989.

M. Horiguchi, M. Aoki, Y. Nakagome, S. Ikenaga, K. Shimohigashi, "An Experimental Large-Capacity Semiconductor File Memory Using 16-Levels/Cell Storage", *IEEE Journal of Solid-State Circuits*, Vol. 23, No. 1, pp. 27–33, February 1988.

C. Hu, "Lucky-Electron Model for Channel Hot-Electron Emission", *1979 IEDM Tech. Dig.*, pp. 22–25, December 1979.

C. Hu, "Future CMOS Scaling and Reliability", *Proceedings of the IEEE*, Vol. 81, pp. 682–689, 1993.

IEEE 1995, Nonvolatile Semiconductor Memory Workshop, "Flash Memory Tutorial", Monterey, California, August 14, 1995.

IEEE Standard Department, "IEEE P1005 draft standard for definitions, symbols, and characteristics of floating gate memory arrays", approved 1998.

D. Ielmini, A.S. Spinelli, A.L. Lacaita, L. Confalonieri, A. Visconti, "New technique for Fast Characterisation of SILC Distribution in Flash Arrays", *Proc. IRPS*, pp. 73–80, 2001.

D. Ielmini, A.S. Spinelli, A.L. Lacaita, R. Leone, A. Visconti, "Localisation of SILC in Flash Memories after Program/Erase Cycling", *Proc. IRPS*, pp. 1–6, 2002.

D. Ielmini, A. Spinelli, A. Lacaita, A. Modelli, "Statistical Model of Reliability and Scaling Projections for Flash Memories", *IEDM Tech. Dig.*, 2001.

T. S. Jung et al., "A 3.3-V 128-Mb Multilevel NAND Flash Memory for Mass Storage Applications", in *ISSCC Dig. Tech. Papers*, pp. 32–33, February 1996.

T. Kawahara et al., "Bit-Line Clamped Sensing Multiplex and Accurate High-Voltage Generator for 0.25 µM Flash Memories", in *1996 IEEE Int. Solid-State Circuits Conf. Dig. Tech. Pap.*, pp. 38–39, February 1996.

S. Keeney, "A 130 nm Generation High-Density ETOX Flash Memory Technology", *IEDM Tech. Dig.*, p. 41, 2001.

S. Kenney et al., "Complete Transient Simulation of Flash EEPROM Devices", *IEEE Transaction on Electron Devices*, Vol. 39, pp. 2750–2757, 1992.

O. Khouri, S. Gregori, R. Micheloni, D. Soltesz, G. Torelli, "Low Output Resistance Charge Pump for Flash Memory Programming", *2001 IEEE Proc. Int. Workshop on Memory Technology, Design and Testing, San Jose, CA (USA)*, pp. 99–104, August 2001.

O. Khouri, S. Gregori, A. Cabrini, R. Micheloni, G. Torelli, "Improved Charge Pump for Flash Memory Applications in Triple-Well CMOS Technology", in *2002 IEEE Proc. Int. Symposium on Industrial Electronics, L'Aquila (Italy)*, pp. 1322–1326, July 2002.

O. Khouri, R. Micheloni, S. Gregori, G. Torelli, "Fast Voltage Regulator for Multilevel Flash Memories", in *Records 2000 IEEE Int. Workshop on Memory Technology, Design and Testing*, pp. 34–38, August 2000.

O. Khouri, R. Micheloni, I. Motta, A. Sacco, G. Torelli, "Capacitive Boosting Circuit for the Regulation of the Word Line Reading Voltage in Non-Volatile Memories", *U.S. Patent* No. 6.259.635.

O. Khouri, R. Micheloni, I. Motta, G. Torelli, "Word-Line Read Voltage Regulator with Capacitive Boosting for Multimegabit Multilevel Flash Memories", in Proc. European Conf. Circuit Theory and Design 1999, Vol. I, pp. 145–148, August–September 1999.

O. Khouri, R. Micheloni, A. Sacco, G. Campardo, G. Torelli, "Program Word-Line Voltage Generator for Multilevel Flash Memories", in *Proc. 7th IEEE Int. Conf. on Electronics, Circuits, and Systems*, Vol. II, pp. 1030–1033, December 2000.

O. Khouri, R. Micheloni, G. Torelli, "Very Fast Recovery Word-Line Voltage Regulator for Multilevel Non-volatile Memories", in *Proc. Third IMACS/IEEE Int. Multiconference Circuits, Communications and Computers*, Athens, Greece, pp. 3781–3784, June 1999.

Y. Konishi et al., "Analysis of Coupling Noise Between Adjacent Bit Lines in Megabit DRAM's", *IEEE Journal of Solid-State Circuits*, Vol. 24, No. 1, February 1989.

V. N. Kynett et al., "An In-System Reprogrammable 256 K CMOS Flash Memory", *ISSCC, Conf. Proc.*, pp. 132–133, 1988.

V. N. Kynett et al., "A 90-ns one-million Erase/Program Cycle 1-Mbit Flash memory", *IEEE Journal of Solid-State Circuits*, Vol. SC-24, pp. 1259–1264, October 1989.

M. Lenzlinger, E. H. Show, "Fowler-Nordheim Tunnelling into Thermally Grown SiO2", IEDM Technical Digest, Vol. 40, pp. 273–283, 1969.

M. Maccarrone et al., "Program Load Adaptive Voltage Regulator for Flash Memories", *Journal of Solid-State Circuit*, Vol. 32, No. 1, p. 100, January 1997.

S. Mahapatra, S. Shukuri, J. Bude, "CHISEL Flash EEPROM–Part I: Performance and Scaling", *IEEE Trans. Electron Devices*, Vol. ED-49, pp. 1296–1301, July 2002.

F. Maloberti, "Analog Design for CMOS VLSI System", *Kluwer Academic Publishers, Boston*, 2001.

T. Mano, J. Yamada, J. Inoue, and S. Nakajima, "Circuit Techniques for a VLSI Memory", *IEEE Journal of Solid-State Circuits*, Vol. 18, No. 5, pp. 463–469, October 1983.

S. Maramatsu et al., "The solution of Over-Erase Problem Controlling Poly-Si Grain Size-Modified Scaling Principles for Flash Memory", *IEDM Tech. Dig.*, pp. 847–850, 1994.

M. McConnel et al. , "An Experimental 4-Mb Flash EEPROM with Sector Erase", *IEEE Journal of Solid-State Circuits*, Vol. 26, No. 4, pp. 484–489, April 1991.

R. Micheloni et al., "Row Decoder Circuit for an Electronic Memory Device, Particularly for Low Voltage Applications", *USA Patent* No. 6,069,837.

R. Michcloni et al., "Line Decoder for a Low Supply Voltage Memory Device", *USA Patent* No. 6,111,809.

R. Micheloni et al., "Nonvolatile Memory Device, in Particular a Flash-EEPROM", *USA Patent* No. 6,351,413.

R. Micheloni et al., "Method and a Related Circuit for Adjusting the Duration of a Synchronization Signal ATD for Timing the Access to a Non-volatile Memory", *USA Patent* No. 6,075,750.

R. Micheloni et al., "Method for Reading a Multilevel Nonvolatile Memory and Multilevel Nonvolatile Memory", *USA Patent* No. 6,301,149.

R. Micheloni et al., "Read Circuit for a Nonvolatile Memory", *USA Patent* No. 6,327,184.

R. Micheloni, M. Zammattio, G. Campardo, O. Khouri, G. Torelli, "Hierarchical Sector Biasing Organization for Flash Memories", in *Records 2000 IEEE Int. Workshop on Memory Technology, Design and Testing*, pp. 29–33, August 2000.

R. Micheloni, I. Motta, O. Khouri, G. Torelli, "Stand-by Low-Power Architecture in a 3-V only 2-bit/cell 64-Mbit Flash Memory", in *Proc. 8th IEEE Int. Conf. Electronics, Circuits, and Systems*, Vol. II, pp. 929–932, September 2001.

R. Micheloni, I. Motta, O. Khouri, G. Torelli, "Stand-by Low-Power Architecture in a 3-V only 2-bit/cell 64-Mbit Flash Memory", in *Proc. 8th IEEE Int. Conf. Electronics, Circuits, and Systems*, Vol. II, pp. 929–932, September 2001.

R. Micheloni et al., "A 0.13-μm CMOS NOR Flash Memory Experimental Chip for 4-b/cell Storage", *ESSCIRC 28th Proc. European Solid-State Circuit Conf.*, pp. 131–134, September 2002.

R. Micheloni et al. , "The Flash Memory Read Path Building Blocks and Critical Aspects", *IEEE Proceeding of the*, Vol. 91, No. 4, pp. 537–553, April 2003.

M. Mihara, Y. Terada, M. Yamada, "Negative Heap Pump for Low Voltage Operation Flash Memory", in *1996 Symposium VLSI Circuits Dig. Tech. Pap.*, pp. 76–77, June 1996.

K. Ming-Dou, W. Chung-Yu, W. Tain-Shun, "Area-Efficient Layout Design for CMOS Output Transistors", *IEEE, Trans. On Electron Devices*, Vol. 44, No. 4, April 1997.

Y. Miyawaki et al., "A New Erasing and Row Decoding Scheme for Low Supply Voltage Operation 16-Mb/64-Mb Flash Memories", *IEEE Journal of Solid State Circuits*, Vol. 27, No. 4, April 1992.

Y. Mochizucki, "Read-Disturb Failure in Flash Memory at Low Field", Intel reports, Nikkei Electronics Asia, pp. 35–36, May 1993.

A. Modelli, R. Bez, A.Visconti "Multi-level Flash Memory Technology", *2001 International Conference on Solid State Devices and Materials (SSDM)*, Tokyo, Extended Abstract, pp. 516–517, 2001.

A. Modelli, A. Manstretta, G. Torelli, "Basic Feasibility Constraints for Multilevel CHE-Programmed Flash Memories", *IEEE Transaction Electron Devices*, Vol. ED-48, pp. 2032–2042, September 2001.

S. Mori et al., "ONO Interpoly Dielectric Scaling for Non-Volatile Memories Applications", *IEEE Trans. On Electron Devices*, Vol. 38, No. 2, pp. 386–391, 1991.

H. Morimura, N. Shibata, "A Step-Down Boosted-Wordline Scheme for 1-V Battery-Operated Fast SRAM's", *IEEE Journal of Solid-State Circuits*, Vol. SC-33, No. 8, August 1998.

I. Motta, G. Ragone, O. Khouri, G. Torelli, R. Micheloni, "High-Voltage Management in Single-Supply CHE NOR-Type Flash Memories", *Proceedings of the IEEE*, Vol. 91, pp. 554–568, 2003.

S. Mukherjee, T. Chang, R. Pang, M. Knecht, D. Hu, "A Single Transistor EEPROM Cell and its Implementation in a 512 K CMOS EEPROM", *IEDM Tech. Dig.*, pp. 616–619, 1958.

K. Natori, "Sensitivity of dynamic MOS Flip-Flop Sense Amplifiers", *IEEE Transaction on Electron Devices*, Vol. ED-33, No. 4, pp. 482–488, April 1986.

M. Ohkawa el al., "A 9.8 mm 2 Die size 3.3 V 64 Mb Flash Memory with FN-NOR type Four Level Cell", *IEEE Journal of Solid-State Circuit*, Vol. 31, No. 11, p. 1584, November 96.

N. Otsuka, M. A. Horowitz, "Circuit Techniques for 1.5-V Power Supply Flash Memory", *IEEE Journal of Solid-State Circuits*, Vol. 32, No. 8, August 1997.

P. Pavan, R. Bez, P. Olivo, E. Zanoni, "Flash Memory Cells – An Overview", *Proceedings of the IEEE*, Vol. 85, pp. 1248–1271, 1997.

A. Pierin, S. Gregori, O. Khouri, R. Micheloni, G. Torelli, "High-Speed Low-Power Sense Comparator for Multilevel Flash Memories" in *Proc. 7th Int. Conf. Electronics, Circuits and Systems,* Vol. II, pp. 759–762, December 2000.

B. Prince, "Semiconductor Memories. A Handbook of Design Manufacture and Application", Wiley & Sons, 1993.

D. B. Ribner, M. A. Copeland, Design Techniques for Cascode CMOS Op Amps with Im-proved PSRR and Common Mode Input Range, *IEEE Journal of Solid-State Circuits*, Vol. SC-19, No. 6, pp. 919–925, December 1984.

B. Riccò et al., "Nonvolatile Multilevel Memories for Digital Applications", *Proc. IEEE*, Vol. 86, pp. 2399–2421, December 1998.

G. A. Rincon-Mora, P. E. Allen, "A Low-Voltage, Low Quiescent Current, Low Drop-out Regulator", *IEEE Journal of Solid-State Circuits*, Vol. SC-33, pp. 36–44, January 1998.

P. L. Rolandi et al., "A 32 Mb-4b/cell Analog Flash Memory Supporting Variable Density with Only Supply and serial I/O", 25th ESSCIRC '99.

P. L. Rolandi et al., "1M-cell 6b/cell Analog Flash Memory for Digital Storage", *IEEE ISSCC Dig. Tech. Papers*, pp. 334–335, February 1998.

T. Sakurai, K. Tamaru, "Single Formulas for Two-and Three-Dimensional Capacitances", *IEEE Transactions on Electron Devices*, Vol. ED-30, No. 2, pp. 183–185, February 1983.

L. Selmi, C. Fiegna, "Physical Aspects of Cell Operation And Reliability", in *Flash Memory*, P. Cappelletti et al., Ed Norwell, Ma: Kluwer, 1999.

A. Silvagni et al., "Modular Architecture For a Family of Multilevel 256/192/128/64Mb 2-Bit/Cell 3V Only NOR FLASH Memory Devices", *ICECS 2001*, Malta, September 2001.

S. M. Sze, "Physics of Semiconductor Device", *John Wiley & Sons, Inc, New York*, 1969.

Y. Tang et al., "Different Dependence of Band-to-Band and Fowler-Nordheim Tunneling on Source Doping Concentration of an n-MOSFET", *IEEE Electron Device Letters*, Vol. 17, p. 525, 1996.

T. Tanzawa, S. Atsumi, "Optimization of Word-Line Booster Circuits for Low-Voltage Flash Memories", *IEEE Journal of Solid-State Circuits*, Vol. SC-34, pp. 1091–1098, August 1999.

T. Tanzawa, T. Tanaka, "A Dynamic Analysis of the Dickson Charge Pump Circuit", *IEEE Journal of Solid-State Circuits*, Vol. SC-32, No. 8, pp. 1231–1240, August 1997.

U. Tietze, CH. Schenk, "Advanced electronic circuit", *Springer-Verlag Berlin Heidelberg New York*, 1978.

A. Umezawa et al., (1992), "A 5 V-only Operation 0.6-um Flash EEPROM with Row Decoder Scheme in Triple-Well Structure", *IEEE Journal of Solid-State Circuits*, Vol. 27, pp. 1540–1546.

S. T. Wang, "On the I-V Characteristics of Floating-Gate Mos Transistors", *IEEE Transaction on Electron Devices*, Vol. ED-26, No. 9, September 1979.

C. C. Wang, J. Wu, "Efficiency Improvement in Charge Pump Circuits", *IEEE Journal of Solid-State Circuits*, Vol. SC-32, pp. 852–860, June 1997.

M. Zammattio, I. Motta, R. Micheloni, C. Golla, "Low Consumption Voltage Boost Device", *U.S. Patent* No. 6.437.636.

M. Zhang, N. Llaser, F. Devos, "Improved Voltage Tripler Structure with Symmetrical Stacking Charge Pump", Electronics Letters, Vol. 37, pp. 668–669, May 2001.

4 NAND Flash memories

4.1 Introduction

The NAND Flash architecture was introduced by Toshiba in 1989. NAND Flash high-density, low power, cost effectiveness, and scalable design make it an ideal choice to fuel the explosion of multimedia products, like USB keys, MP3 players and digital cameras.

Due to the efficient architecture of the NAND Flash, its cell size is, by a matter of fact, almost half the size of a NOR cell. This enables NAND Flash architecture to offer higher densities with larger capacity on a given die size, in combination with a simpler production process. With the last generation of NAND technologies (below 50 nm) it is possible to integrate on the same piece of silicon 2 GB, i.e. half of a DVD. Having a system with some tens of these memories, immediately gives the possibility of a storage system (solid-state disk) competitive with a traditional hard disk drive.

Advances in system design techniques also enable the more cost effective NAND Flash to replace NOR Flash in a significant percentage of traditionally NOR Flash applications.

As in the previous chapter, we now proceed with a description of the basic functionalities of a NAND memory.

4.2 Read

The architecture of NAND type Flash memories requires the cells to be connected in series, in groups constituted by 16, 32 or 64 cells. In series with the cells there are two more selection transistors that connect the series of the cells to the *Source Line* or to the bitlines even (BLe) and odd (BLo). This basic structure is illustrated in Fig. 4.1.

When we read a cell, its gate is fixed at 0 V, while the other cells are biased with high voltages (usually 4÷5 V), so that they operate as pass-transistors, independently from the value of their threshold voltages.

An erased Flash cell has a threshold voltage smaller than 0 V, vice versa, if written, its threshold voltage is positive but, however, smaller than 4 V. In practice, biasing the gate of the selected cell with a voltage equal to 0 V, the series of all the cells will conduct current if the addressed cell is erased, vice versa it will not conduct current if the cell is written.

In Fig. 4.2 we sketch the V_T threshold voltage distributions for written and erased cells. Note that, for gate voltages higher than the right margin (V_{TWR}) of the written distribution, the cell always conducts current whatever its threshold is: this characteristic is exploited exactly when the cell has to operate as a pass-transistor.

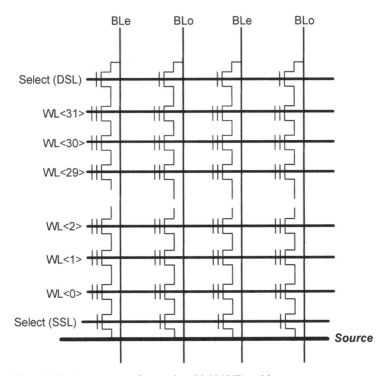

Fig. 4.1. Basic structure of a matrix with NAND architecture

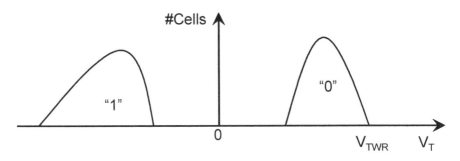

Fig. 4.2. Threshold voltage distributions for written and erased cells

Unlike NOR type Flash memories, the current to be detected in these structures, with different cells in series, is very low. Such current value is usually $200\div300$ nA (in NOR memories we speak of about tens of μA). It is unthinkable to try to detect such current with the NOR differential structure described in the preceding chapter.

The read mode of NAND memories is a charge integration mode, exploiting the bitline parasitic capacitance. This capacitance is precharged at a fixed value (usually 1.2 V): if the cell is erased and conducts current the capacitance discharges, otherwise, if the cell is written, it does not conduct current and the bitline will keep its initial charge value. Different circuits exist to detect if the bitline parasitic capacitance has been discharged: the structure depicted in Fig. 4.3 is present in almost all solutions. The bitline parasitic capacitance is indicated with C_{BL} while the NAND string (Fig. 4.1) is equivalent to a current generator.

During the charge of the bitline, the gate of the PMOS transistor M_P is kept grounded, while the gate of the NMOS transistor M_N is kept at a fixed value V_1. Typical value for V_1 is around 2 V. At the end of the charge transient the bitline will have a voltage V_{BL}:

$$V_{BL} = V_1 - V_{THN} \qquad (4.1)$$

where V_{THN} indicates the threshold voltage value of the NMSO M_N.

This phase of bitlines precharge is very demanding in terms of power consumption from the external power supply VDD. Today we are talking of NAND arrays with more than 32 K bitlines and all of them have to be precharged. Keep in mind that the parasitic capacitance of each bitline can be up to $2\div3$ pF. In order to reduce the current peaks from VDD, the charge of the parasitic capacitances is controlled over the time and, generally, lasts in the order of $2\div6$ μs. The OUT node, instead, is charged at the VDD supply voltage. At this point, the M_N and M_P transistors are switched off and the OUT and BL nodes are left floating, precharged at their initial voltage values.

Now starts the real *evaluation phase* of the current of the cell: it consists in verifying whether the cell absorbs current or not. If the cell does not absorb current, the bitline capacitance remains charged at its initial value, otherwise the bitline starts to discharge. At the end of the T_{VAL} time (we will see what its value is and on which factors it depends in a while) the gate of the NMOS transistor M_N is biased to a value $V_2<V_1$, usually $1.6\div1.4$ V.

In the case a T_{VAL} time has elapsed long enough to discharge the bitline voltage under the value:

$$V_{BL} < V_2 - V_{THN} \qquad (4.2)$$

the M_N transistor "turns on" and the voltage of the OUT node becomes equal to the one of the bitline.

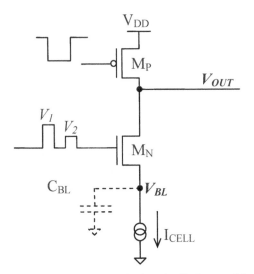

Fig. 4.3. Structure to calculate the discharge of the bitline and

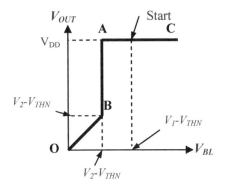

Fig. 4.4. Characteristic $V_{OUT} - V_{BL}$ of the circuit reported in Fig. 4.3

Figure 4.4 shows the relationship between the voltage of the OUT node and V_{BL} when the V_2 voltage is applied to the gate of the M_N transistor: the point indicated with "Start" corresponds to the end of the precharge phase of the bitline and, therefore, to the beginning of the evaluation phase. The segment indicated with AC corresponds either to the cell which does not absorb current or to a cell which does not sink enough current to discharge the bitline below the value $(V_2 - V_{THN})$

in a T_{VAL} time. Instead, if the cell has brought the bitline to a value lower than $(V_2 - V_{THN})$, then OUT is connected to the bitline and they both have the same voltage (segment OB in Fig. 4.4). Obviously, the transition between the points A and B is not a step; in reality there is a certain slope, which depends on the ratio between the bitline capacitance and the OUT node capacitance (such ratio is in the order of $30 \div 100$).

Let's now go back to the T_{VAL} time: what is the minimum time we have to wait in order to discharge the bitline? This time depends on the value of the bitline capacitance, on the value of the minimum current absorbed by the cell and on the difference between the voltages V_1 and V_2, according to Eq. (4.3).

$$T_{VAL} = \frac{C_{BL}(V_1 - V_2)}{I_{CELL}} \qquad (4.3)$$

Typical values of the bitline capacitance are $2 \div 3$ pF, while the cell can manage to absorb currents in the order of 200 nA. The difference between V_1 and V_2 is about $300 \div 500$ mV. It follows that the evaluation time ranges between 5 and 10 μs. Since the bitline capacitance and the current of the cell cannot be modified too much (they depend on the manufacturing technology used), we could consider decreasing the difference between V_1 and V_2 in order to decrease the evaluation time. In reality, the difference between the two voltages is required to cover up the unwanted effects during evaluation, such as disturbs on the bitline voltages or absorptions of spurious currents overlapping that of the cell (leakage current), which may invalidate the reading.

The output voltage value is "digitalized and frozen" in volatile memory structures made with simple latches. A circuital implementation with latches is illustrated in Fig. 4.5.

At the beginning of the read phase, the DATA_N node is forced to ground through the M_{SE} transistor. The transistors M_N and M_P are the same of Fig. 4.3. At the end of the evaluation phase, the V_{OUT} voltage may be either at VDD (written cell) or at a voltage value corresponding to the segment OB of Fig. 4.4 (erased cell). V_{OUT} is also used to bias the gate of the M_{SE} transistor.

Let's now see what happens if we bring the READ signal to VDD. In the case V_{OUT} is equal to VDD, the M_{SE}-M_{EN} series conducts current and brings the potential of the DATA node to ground: accordingly the voltage of DATA_N node moves to VDD. Vice versa, if OUT has been discharged, the M_{SE}-M_{EN} series is not able to unbalance the latch, which, therefore, remains at its initial state.

In conclusion, the differences between NOR and NAND, force us to use different read techniques: the differential one for NOR memories and the charge integration one for NAND memories. These two methods have different timing: few tens of nanoseconds for the NOR ones and ten microseconds for the NAND ones.

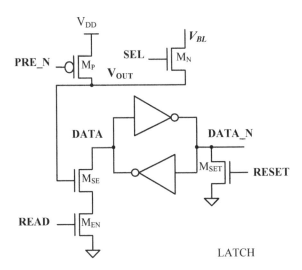

Fig. 4.5. Read circuit (Page Buffer) for a NAND Flash memory

4.3 Program

As already anticipated, the program of NAND memories takes place using a different physical principle compared to NOR: a NAND cell exploits, in fact, the quantum-effect of electron tunneling in the presence of a strong electric field (Fowler-Nordheim tunneling). In particular, depending on the polarity of the electric field applied we obtain the program or erase of the memory cell. The tunneling effects can be of two types.

- *Channel tunneling*: the electric field is applied between the gate and the substrate, while the drain and source terminals are floating; in this case the accumulation of electrons in the floating gate takes place thanks to the transition of the same electrons from the substrate to the gate.
- *Junction tunneling*: in this case the electric field is applied between the floating gate and one of the two terminals (drain or source).

The tunneling used in NAND memories is the channel type.

During program, the number of electrons which cross the oxide is linked to the electric field: in fact the greater such field is, the greater the injection probability. Thus, in order to improve the program performances, it is essential to have high electric fields available and therefore high voltages. This requirement is one of the main drawbacks of this program method, since the oxide degradation is linked to these voltages.

To overcome this problem we could consider a reduction in the dielectric layer. In this way the injection efficiency is improved while the voltages required to obtain the program and the erase of the cell (and therefore the energy of the electrons which cross the oxide) is reduced. Unfortunately the oxide cannot be reduced in thickness indefinitely since its decrease involves other degradation effects, such as the generation of defects which create low field leakage current: the SILC (Stress Induced Leakage Current). Another disadvantage of the tunneling mechanism is linked to the time taken by the cell to program itself, usually longer compared to the time taken by a cell which is programmed through the hot electrons mechanism. The main advantage, instead, is linked to the current required, which is definitely moderate (we speak of nanoAmpere per cell). This is what makes the Fowler-Nordheim mechanism suitable for a parallel program of an elevated number of cells.

The algorithm used to program the cells of a NAND memory is a *Program & Verify* algorithm (we program and verify if the cell has reached the required "position"), exactly like in the case of NOR memories. What differentiates verify from read is the value with which the addressed cell gate is biased: generally higher during verify in order to have a margin and to guarantee the correct allocation of the distributions. The time, instead, coincides exactly with the read time (see previous section).

In a NAND memory a cell belongs to a string of cells; the string itself is selectable through the opening of the drain and source selectors (Fig. 4.1). In order to trigger the injection of electrons into the floating gate we need the following voltages inside the string:

- the gate of the drain selector at VDD;
- the gate of the cells not involved in the program operation at a voltage between 8 V and 10 V;
- the source selector gate grounded.

In this way, if we now set 0 V on the bitline connected through the DSL drain selector, the cell to be programmed turns out to have 0 V on its drain, while the source is floating and a voltage of 15÷20 V is applied to its gate.

However, at this point, it is interesting to understand how we can prevent the cells sharing the same gate with the programmed one from undergoing an undesired program. We could, for example, put a high voltage on the drains of the cells not to be programmed: in this way the channel voltages of these cells rise, thus reducing the probability of electron injection into the floating gate. This method is however very difficult in practice, if we want to make memories with commercially interesting sizes. For example, for a 32 Mbit memory we could have a matrix organized in 8 k (8128) rows and 4 k (4096) columns. In the hypothesis of programming 2048 cells, we should bring the drains of all the cells sharing the same wordline (in this case 2048 cells) to a high voltage.

The drawbacks of this method are evident:

- a greater area occupied because of the need to have charge pumps able to provide current and voltage to carry out the inhibit of the non-programmed cells;
- sensing circuits able to manage high voltages;
- time needed for the charge of all the bitlines.

In alternative, we can use the so-called *self-boosting* mechanism, where the idea is that of exploiting exactly the high potentials involved during program through the parasitic capacitances of the cell, in order to increase the channel potential itself (Fig. 4.6).

In particular we now introduce the following definitions:

- C_{ONO} is the oxide capacitance between control gate – floating gate;
- C_{tun} is the tunneling oxide capacitance;
- C_{ch} is the channel capacitance.

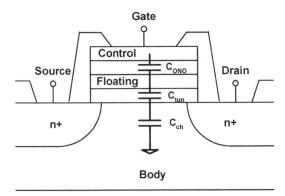

Fig. 4.6. Capacitances contributing to the channel boosting

The bitlines of the unselected cells are precharged at a V_{ini} potential and then left floating. When the unselected wordlines move to the V_{pass} potential, the channel of the cells belonging to the unselected bitline is boosted to the V_{sl} voltage:

$$V_{sl} = V_{ini} + V_{pass} \frac{C_{pl}}{C_{pl} + C_{ch}} \qquad (4.4)$$

where C_{pl} is the equivalent of the tunnel oxide capacitance in parallel with the ONO's one:

$$C_{pl} = \frac{C_{tun} \cdot C_{ONO}}{C_{tun} + C_{ONO}} \qquad (4.5)$$

After the explanation of this mechanism, it is useful to reconsider the voltages applied to the selected and unselected strings (Fig. 4.7).

It is interesting to note that the bitline of the selected cell is biased at 0 V, even if, at the beginning of this section, we stated that the drain and source terminals can be left floating. The bitline biased at 0 V is necessary because we want to prevent couplings with the neighbor bitlines: if the drain of the selected cell rises, the program efficiency is reduced.

The gate of drain sector is instead biased to the supply voltage. That means that the unselected bitlines precharge up to (VDD − V_{THN}), where V_{THN} is the threshold voltage of the drain selector transistor. We could also consider a technique in which the voltage on the drain selector gate is as high as needed to bias the channel exactly at VDD. It is actually a compromise between the need to have a high voltage on the channel in order to improve its boost and the need to contain consumption and time. The gates of the string cells are brought to a voltage between 8 V and 10 V. This value is linked to the boost: the higher this voltage is and the longer it is possible to maintain the channel boost. Nevertheless the choice of this voltage is definitely crucial: a too high value would increase the possibility for cells belonging to the same string of the selected one to self-program (*Vpass Stress*); a too low value could not guarantee the boost for the entire length of the program algorithm, bringing the cells having the gate in common with the selected one to self-program (*Vprogram Stress*).

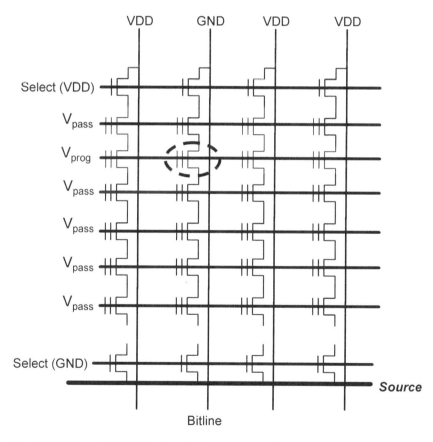

Fig. 4.7. Biasing of a NAND string during program. The cell to be programmed is the highlighted one

The source selector is kept off, while the string source is biased at VDD. This choice is made to have the source – channel junction (the body is kept to ground for obvious reasons) biased in an inversed way, in order to prevent possible leakage currents from passing in the source, thus discharging the boosted channel of the cells not to be programmed.

A further stratagem which can be used to improve the channel boost is the biasing at 0 V of the wordlines (WL) adjacent to the one at the program voltage. For example, if WL<7> is the one biased at the program voltage, the WL<6> and WL<8> are brought to 0 V, the others to the voltage of 8÷10 V. In this way, the channel of the cell to be programmed is better isolated from the other cells and the voltage that it manages to reach is more elevated. This technique is known as *Local Self Boosting*.

It is now possible to say something about the sequence that has to be followed when applying the potentials to program a cell.

First, the bitlines are charged, and so the gate of the selector is brought to VDD. Once this phase is finished, all the wordlines pass from 0 V to the voltage of 8÷10 V. Finally, the wordline to be programmed is raised from 8÷10 V to the program voltage. This transition can take place in two ways:

- the gate of the cell to be programmed is directly brought to the program voltage, letting the parasitic *RC* of the WL determine the charge time;
- the final voltage is reached in a controlled way by imposing a well precise gradient to the voltage, for example 1 V/μs.

In this phase, the gate of the drain selector is either biased at a lower voltage (at about 1.3 V) or, vice versa, left at its initial voltage, depending on how we choose to bias the gate of the cell to be programmed. In fact, due to the rather reduced pitch of the cells, the coupling capacitance between wordlines and selectors gates (but also between the wordlines themselves) is elevated. If the gate of the selected cell is biased directly at the program voltage, we might have, especially when we intend to program the cell adjacent to the drain selector, an increase in the gate voltage of the selector. Since the unselected bitlines are left at VDD, an increase in the gate of the selector could cause a channel boost loss and a consequent program of the cells which are on the same wordline as the programmed one. This effect does not occur if:

- we lower the gate voltage of the selector as soon as the wordlines begin to rise;
- we measure the maximum gradient that triggers the boost on the gate of the selector and we set the program voltage at a slightly inferior gradient.

The same problem appears when the cell nearest to the source selector is programmed. In this case, while the gate of the selector is at GND, the source (which is also the matrix source) is brought to VDD: in this way for the selector to turn on we need to reach the voltage VDD + V_{TH} (its threshold voltage).

Finally, it is important to underline that the program operation in NAND memories, unlike NOR memories, follows a well precise sequence: it starts from the cell nearest to the source selector and it continues along the string until it reaches the cell nearest to the drain selector. This stratagem is necessary for the phenomenon of *Background Pattern Dependency*, since the threshold voltage of a cell depends on the state of the cells situated between the considered cell and the source contact. We already know that the series resistance of a cell is different depending on whether it is programmed or not. If we did not follow a given sequence, it might occur that the threshold voltage of a cell during the verify phase after the program operation might be different from the one during read, simply because other cells below have been programmed in a second time.

4.4 Erase

In Flash memories with NAND architecture it is more convenient to apply the erase pulse by biasing the iP-well with a high voltage and keeping grounded the wordlines of the sector to be erased. What is the reason? In order to generate negative voltages we need negative boosters and, therefore, high-voltage transistors with the triple well option (see Sect. 3.4). Such transistors would be necessary also to build the structures used to bring the negative voltage towards the row decoding; besides, the same row decoding should have at least one high-voltage in triple well transistor to pass the negative voltage to the wordlines. As a matter of fact, avoiding the use of negative voltages means a reduction in lithographic masks and technological complexity.

We also need to consider that, as far as design and layout are concerned, it is very difficult to fit the row decoder into the small pitch of the memory array. Remember that each wordline needs its own driver, made up by high voltage transistors. In NAND memories the pitch is even more reduced than in the NOR case, therefore, it is more convenient to carry out a decoding with only one n-channel transistor to bring voltages greater than or equal to zero. Besides, as we will explain later, the information needed regarding the position of the erased distribution can be obtained with a sufficient precision even using only positive voltages: not even from this point of view there is an incentive for the on-chip production of negative voltages.

In memories with NAND architecture, therefore, the erase pulse is applied to the substrate terminal, which must be brought to a more elevated voltage compared to the NOR case (the common source is left floating). Besides, the source terminal is shared by all the blocks: in this way the matrix is more compact and the multiplicity of the structures which bias the iP-well is drastically reduced (in NOR memories these structures are repeated for each sector or group of sectors). On the contrary, the parasitic capacitance to be loaded is much greater and the sectors not concerned with erase (that is all but one) need to be managed in an appropriate way in order to avoid spurious erasures.

The advantage of the NAND architecture from the erase point of view consists in the fact that it is also possible to use the threshold voltages below zero. In contrast with what happens in the NOR architecture, there are no side effects in bringing the threshold voltages of the cells to a negative level: the read mechanism used in the NAND architecture is based exactly on the fact that the cells biased at GND are able to act as pass transistors. No spurious contribution can be brought by the sub-threshold leakage. When a string is read, the selectors of the addressed string are on, while those of the un-addressed strings are off, thus inhibiting the current flow coming from the other strings (eventually depleted) of the same bitline. It is therefore clear that it is not necessary to have negative voltages in order to switch off the wordlines.

The fact that the threshold voltage distribution of the erased cells is very wide does not have a great influence on the behaviour of the string, in terms of series resistance in read, when all the cells of the string, except the addressed one, are biased with a positive gate voltage (Vpass) of about $4 \div 5$ V.

As NAND memories usually do not require a great precision in the positioning of the erased distribution, the erase operation is usually performed with only one pulse calibrated in such a way to bring the distribution even very negative, followed by a verify phase. In this case the NAND specification is helpful, since it allows the management of the *bad blocks*: if the sector does not pass the verify following the erase pulse, it must be considered fail by the customer, who is expected to store this information and not to use the bad sector anymore (whereas if the malfunction occurs during the factory test, the sector can directly be marked as failed). When using the memory, the customer has to check all the sectors in order to avoid the ones which are already fail.

The way the erase verify is performed in the NAND architecture is substantially different from the one used in the NOR architecture. Having no negative voltages available, it is not possible to know the borders of the erased distribution exactly. On the other hand, it is important that all the cells of the string, after the erase operation, can be read as erased: that is, they must be read as logic "1" when their gate is grounded and the other cells of the string are biased as pass transistors. This condition has to apply to all the cells of the string.

The erase verify in the NAND architecture can, therefore, be translated into a requirement for the string, since, biasing all wordlines at GND, it still requires the selected string to be read as erased. As already said, this does not allow us to have a precise information on the position of the erased distribution, but it makes us save a lot of time during the verify phase. In fact, with only one reading, it is now possible to have the status of the whole string, instead of performing as many read operations as the number of cells composing it.

A further requirement to be satisfied is that the cells need to be erased with a margin high enough to compensate for the cycling degradation (Chap. 5). Figure 4.8 shows an example of the cycling window of a NAND cell as the number of program/erase cycles increases, where program and erase are performed without verify. We can note that the threshold of both the erased cell and the programmed cell increases as the number of cycles increases. The effect is due to the phenomena of gain degradation and charge trapped in the oxide. The macroscopic effect on the erase operation is that the pulse which at time zero allows us to satisfy the erase verify, could be insufficient as the number of cycles increases.

Cycling degradation is also present in the NOR architecture; in that case, however, the voltages are calibrated so that the cells are erased at time zero without applying the maximum voltage available. As a consequence of cycling degradation, the erase time increases as more erase pulses are needed together with the soft-program operation (see Sect. 3.4). NOR memories specifications are built in a way to handle this situation.

In the NAND architecture, instead, the specification does not leave room for a second erase pulse: for this reason the calibration of the erase pulse is the result of accurate analyses at process level.

On the other hand, we have to say that technology below 50 nm will require a relaxation of the erase specifications also for NAND.

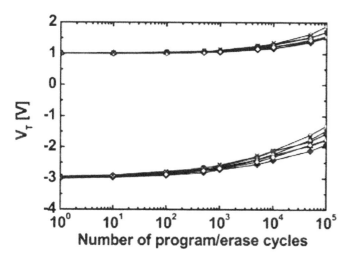

Fig. 4.8. Typical cycling window for NAND Flash memories

Also in the NAND architecture it may be useful to make a preconditioning operation (see Sect. 3.4) before the electrical erase: a more uniform cycling over the entire population of the block helps in containing the distribution spread. Thanks to the fact that also the program operation occurs via tunneling, the *program all 0* can be performed simultaneously on all the rows and all the bitlines of the block, saving a lot of time.

The erase sequence of the NAND architecture is shown in Fig. 4.9, together with the effect the various phases have on the distribution. The program all 0 is not shown because it is not frequently used (and, anyway, its effect is analogous to that of the NOR architecture).

The nodes involved in the erase operation must be discharged carefully also in the case of NAND architecture. The bitlines are floating and, therefore, they can reach a high potential which depends on the substrate voltage and on the capacitive loads involved. The substrate discharge must be carried out in a controlled way, as it is done in NOR memories. The source node can initially be left floating (it will start to discharge together with the substrate by coupling), to be then connected to ground once the substrate discharge has finished. Also the bitlines start to discharge together with the substrate by coupling, and are then really discharged to ground through specific transistors.

On the side of the row decoder (Fig. 4.10), the selected wordlines are already kept grounded but those not selected are floating, therefore they are discharged to ground by activating the whole row decoder.

On a technological and manufacturing level it is necessary to pay great attention to the electric characteristics of the row decoding transistors, whose leakage has to be the lowest possible: the rows of the un-addressed sectors are left floating during the erase pulse, and therefore free to charge with the substrate by means of capacitive coupling, so as to allow the tunneling. However, if there is a leakage on the row decoding, the un-addressed wordlines may discharge; if that occurs, the potentials at the ends of the tunnel oxide may become enough to trigger a spurious erase of the unselected blocks.

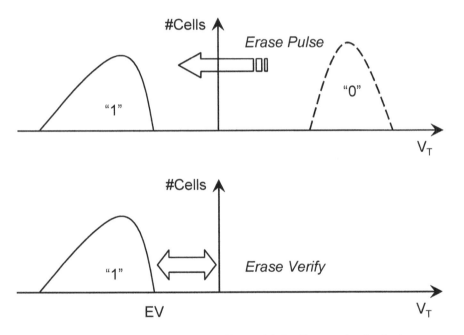

Fig. 4.9. The phases of a NAND erase algorithm and their effects on the distribution

As a summary, in a NAND Flash the typical erase time of a block is about 1 ms, divided between the electrical erase pulse, which lasts about 800 μs, and the erase verify phase of about 100 μs. The program all 0 is usually not executed; it should be, anyway, not longer than 100 μs.

Note that we have not mentioned the size of the block: unlike what happens in the NOR architecture, in the NAND algorithm the erase time of a block is really independent from the number of memory cells and wordlines. In fact, none of the operations described so far is carried out either on a page basis or on a wordline basis.

On the other hand, it is clear that shrinking down the technology will require more sophisticated erase algorithms also for NAND Flash memories. As the reliability margins are usually shrinking with the technology node, we can expect that, in the near future, a program after erase (see Sect. 5.2.3) will be introduce in order to contain the erase distribution width.

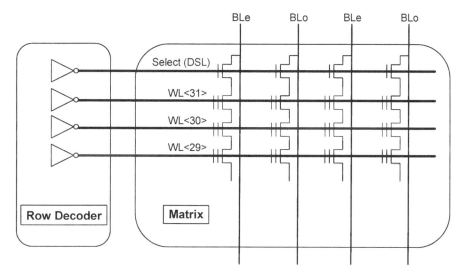

Fig. 4.10. Row decoder is the circuit used to transfer the desired voltage on each wordline

Bibliography

D. S. Byeon et al, "An 8Gb Multi-level NAND Flash Memory with 63 nm STI CMOS Process technology", ISSCC 2005 Digest of Technical Papers, pp. 46–47.

T. Cho et al., "A 3.3 V 1 Gb Multi-level NAND Flash Memory with Non-uniform Threshold Voltage Distribution", ISSCC 2001 Digest of Technical Papers, pp. 28–29.

Y.-J. Choi, K.-D. Suh, Y.-N. Koh, J.-W. Park, K.-J. Lee, Y.-J. Cho, B.-H. Suh, "A High Speed Programming Scheme for Multi-Level NAND Flash Memory", in 1996 Symposium VLSI Circuits Dig. Tech. Papers., pp. 170–171, June 1996.

T. Cho, Y.-T. Lee, E.-C. Kim, J.-W. Lee, S. Choi, S. Lee, D.-H. Kim, W.-G. Han, Y.-H. Lim, J.-D. Lee, J.-D. Choi, and K.-D. Suh. "A Dual-Mode NAND Flash Memory: 1-Gb Multilevel and High-Performance 512-Mb Single-Level Modes," *IEEE Journal of Solid-State Circuits*, Vol. 26, No. 11, pp. 1700–1706, November 2001.

T. Endoh et al., "A New Write/Erase Method to Improve the Read Disturb Characteristics Based on the Decay Phenomena of Stress Leakage Current for Flash Memories", *IEEE Transactions on Electron Devices*, Vol. 45, No. 1, pp. 98–104, January 1998.

T. Hara et al, "A 146mm/sup 2/8 Gb NAND Flash Memory with 70 nm CMOS Technology", ISSCC 2005 Digest of Technical Papers, pp. 44–45.

G. J. Hemink, T. Tanaka, T. Endoh, S. Aritome, R. Shirota, "Fast and Accurate Pro-gramming Method for Multi-Level NAND EEPROMs", in 1995 Symposium VLSI Technology Dig. Tech. Papers., pp. 129–130, June 1995.

K. Hosono et al., "A High Speed Failure Bit Counter for the Pseudo Pass Scheme (PPS) in Program Operation for Giga Bit NAND Flash", NVMWS 2004.

K. Inamiya et al., "A 130 mm2 256 Mb NAND flash with shallow trench isolation technology", ISSCC 1999 Digest of Technical Papers, pp. 112–113.

T. S. Jung, Y.-J. Choi, K.-D. Suh, B.-H. Suh, J.-K. Kim, Y.-H. Lim, Y.-N. Koh, J.-W. Park, K.-J. Lee, J.-H. Park, K.-T. Park, J.-R. Kim, J.-H. Lee, H.-K. Lim , "A 117-mm^2 3.3 V only 128-Mb Multilevel NAND Flash Memory for Mass Storage Applications", *IEEE Journal of Solid-State Circuits*, Vol. SC-31, pp. 1575–1583, November 1996.

K. Kim, "Future Outlook of NAND Flash Technology for 40 nm Node and Beyond", NVMWS 2004.

J. -D. Lee, "Effects of Floating-Gate Interference on NAND Flash Memory Cell Operation", *IEEE Electron Device Letters*, Vol. 23, No. 5, pp. 264–266, May 2002.

S. Lee et al, "A 3.3 V 4 Gb Four-level NAND Flash Memory with 90 nm CMOS technology", ISSCC 2004 Digest of Technical Papers, pp. 52–53.

J. Lee et al, "A 1.8 V 2 Gb NAND flash memory for mass storage applications" ISSCC 2003 Digest of Technical Papers, pp. 290–494.

F. Masuoka, M. Momodomi, Y. Iwata and R. Shirota, "New Ultra High Density EPROM and Flash with NAND Structure Cell", IEDM Tech. Dig., pp. 552–555, 1987.

R. Micheloni et al. "A 4Gb 2b/Cell NAND Flash Memory with Embedded 5b BCH ECC for 36MB/s System Read Throughput", ISSCC Dig. Tech. Papers, San Francisco, February 2006.

H. Nobukata et al., "A 144-Mb, Eight-Level NAND Flash Memory with Optimized Pulsewidth Programming," *IEEE Journal of Solid-State Circuits*, Vol. 35, No. 5, pp. 682–690, May 2000.

K. Sakui, "NAND Flash design" 2004 ISSCC, Memory Forum, Memory Circuit Design: Non-Volatile Memories Technology and Design.

K. -D. Suh et al., "A 3.3 V 32 Mb NAND Flash Memory with Incremental Step Pulse Programming Scheme," *IEEE Journal of Solid-State Circuits*, Vol. 30, No. 11, pp. 1149–1156, November 1995.

K. Takeuchi et al., "A 56 nm CMOS 99 mm^2 8 Gb Multi-level NAND Flash Memory with 10MB/s Program Throughput" 2006 ISSCC Digest of Technical Papers, pp. 507–508.

K. Takeuchi, T. Tanaka, H. Nakamura, "A Double-Level-Vth Select Gate Array Architecture for Multilevel NAND Flash Memories", *IEEE Journal of Solid-State Circuits*, Vol. SC-31, pp. 602–609, April 1996.

K. Takeuchi, T. Tanaka, T. Tanzawa, "A Multipage Cell Architecture for High-Speed Programming Multilevel NAND Flash Memories," *IEEE Journal of Solid-State Circuits*, Vol. 33, No. 8, pp. 1228–1238, August 1998.

K. Takeuchi et al., "A Negative Vth Cell Architecture for Highly Scalable, Excellent Noise Immune and High Reliable NAND Flash Memories," *IEEE Journal of Solid-State Circuits*, Vol. 34, No. 5, pp. 675–684, May 1999.

K. Takeuchi et al., "A Source-Line Programming Scheme for Low-Voltage Operation NAND Flash Memories," *IEEE Journal of Solid-State Circuits*, Vol. 35, No. 5, pp. 672–681, May 2000.

T. Tanaka et al., "A quick intelligent page-programming architecture and a shielded bitline sensing method for 3V-only NAND flash memory", *IEEE Journal of Solid-State Circuits*, Vol. 29, No. 11, November 1994, pp. 1366–1373.

T. Tanaka, T. Tanzawa, K. Takekuchi, "A 3.4-Mbyte/sec Programming 3-level NAND Flash Memory Saving 40% Die Size per Bit", in 1997 Symposium VLSI Circuits Dig. Tech. Papers., pp. 65-66, June 1997.

5 Reliability of floating gate memories

Andrea Chimenton[a], Massimo Atti[b], Piero Olivo[a]

[a]Dipartimento di Ingegneria
Università di Ferrara
Via Saragat 1
44100, Ferrara
[b]Infineon Technologies Italia S.r.l.
Automotive, Industrial and Multimarket
Microcontroller Product Engineering
Via Niccolò Tommaseo, 65
35131, Padova

5.1 Introduction

The most recent achievement of floating gate storage is a 8 Gb Multi-Level NAND cell developed in 63 nm technology with a cell size of 0.0164 μm^2 that well fits the predictions of ITRS roadmap (Park et al. 2004). The design and production of such a high cell density is a tough challenge to be overcome by engineers. In particular, reliability requirements are becoming more and more difficult to guarantee moving towards the limits of the technology. The floating gate concept is predicted to face technological limits beyond the 40 nm node, even if the limits for NAND cells could be pushed a little further than for NOR ones. The main physical limits that prevent further scaling of the cells are: cell to cell interference, due to the parasitic capacitive coupling among neighboring floating gates; low coupling ratio with the control gate, which results also in a small stored charge (Kim and Choi 2006). On the other side, trap-related leakage currents in the dielectrics prevent any further scaling of the cell dielectrics, which could relieve some of the issues.

In traditional 2-level non-volatile memories (NVM) the improvement of fabrication yield has always relied on a strategic use of column and/or row redundancy (Cappelletti et al. 1999). Soft errors due to alpha particles or other reliability issues were not a serious concern. As a consequence, on-chip ECC techniques have seldom been adopted so far, although their use has been proposed in some cases to improve data retention (i.e. the ability to maintain and retrieve the stored information) and endurance (i.e. the allowed number of repeated program/erase cycles) or to enhance both yield and reliability (Nakayama et al. 1989; Vancu et al. 1990;

Tanzawa et al. 1997). More commonly, off-chip ECC techniques have been used for NAND-type Flash memories to increase reliability at the board level against the generation of point defects during the programming cycles.

A larger interest in ECC techniques can be found in the embedded Flash memory market, where ECC is widely used in complex System on Chips (SoC) for Automotive applications. In this market it is mandatory to ensure very low defect rate over the whole lifetime (typically 10 years or more). Nowadays Automotive applications require 0.1 ppm (parts per million), or even lower, defects. At the same time SoCs must ensure a quite high data throughput, either for code fetch or for data access from the system memories (including SRAMs, Flash memories, ROMs). This leads to complex architectures including first level cache memory. Data consistency must be ensured on the whole data-path, from the flash memory to the cache and CPU. ECC encoder-decoder circuits could be therefore placed on chip, at different levels: inside the flash memory macro (or at macro's boundary), at the bus memory unit or at the CPU data bus ports. The first solution is more viable when a flash memory has to be covered by ECC to achieve a target level of defects. The other solutions have to be selected if some, or all, memories have to be covered by ECC. In any case, the ECC encoder/decoder circuit has to be inserted in the critical data path, and therefore it introduces an additional delay (the circuit is a combinatory design) that typically enlarges the number of memory access wait-states. Both wider data bus and flash memory burst-access mode have to be considered in order to minimize the throughput penalty. In addition ECC usage adds area penalty in terms of both memory size and additional logic which can have a large impact on the economical point of view.

ECC is also crucial for multilevel (ML) Flash memories that can store more bits per cell. ML has recently been introduced to increase the memory capacity (total number of bits per chip) without significantly reducing the cell size. The use of ML storage has its main drawbacks in reduced speed, reduced signal-to-noise ratio and lower error immunity. This last issue can be solved by design with ECC and can be paid in terms of a further increase in the access time.

In new generation, high density, scaled Flash memories (embedded and especially ML Flash), reliability issues are becoming more and more critical (Aritome et al. 1993; Pavan et al. 1997; Brown and Brewer 1998), thus requiring a more intensive and generalized use of ECC.

This chapter aims at giving an overview of the most important reliability issues affecting the floating gate memory technology which are the key elements whose evaluation is on the basis of each ECC strategy.

Reliability issues will be discussed by considering the single cell first (intrinsic cell degradation) and then by introducing all the important effects derived from considering a large array of cells. As it will be shown, many parasitic physical mechanisms can show up in a scaled floating gate cell, but their actual impact on reliability can be different depending on the array architecture.

The main reliability issues that will be discussed are: intrinsic oxide degradation, data retention, endurance, tail bit, fast bits and erratic bits, read and program disturbs.

Any single cell is subject to extremely high electric fields required to move electrons into and from the floating gate during program/erase operations. Charge transport through a thin oxide layer (~ 9 nm) and charge trapping within the oxide provokes an *intrinsic oxide degradation* (oxide aging) with a long term degradation of both *endurance* and *data retention*. Endurance may be affected by the so-called window closure, i.e. the increasing difficulty in modifying the state of a cell caused by electron trapping. Data retention, representing the ability of freezing the charge within the floating gate and therefore of keeping the programmed state, can be affected by charge leakage from the floating gate through the surrounding oxides. Data retention depends directly on the amount of charge stored within the floating gate and on the acceptable charge loss: it is obvious that both cell shrinking and multi-level solutions make the goal of achieving the required data retention a leading challenge.

Besides the reliability effects related to the single cell there are also many reliability issues which are consequence of the cell array, i.e. the effects of a large number of cells considered as a whole and the way they are interconnected together. The former can depend on the unavoidable technological *parameter dispersion* which can impact the overall reliability. The latter determines the choice of the specific writing physical mechanism (and associated reliability issues) and the way the cells can disturb each other (*program and read disturbs*).

Due to the technological parameter dispersion, each state (or level) of the memory is actually associated to a distribution of values of the characteristic state memory parameter (i.e. threshold voltage in the case of floating gate memories). Ambiguous states can be associated to cells potentially belonging to two adjacent distributions. Therefore, in order to avoid inconsistent readings, a proper separation between the distributions, i.e. *a read margin*, has to be guaranteed. The increase of memory density is accompanied by an increase of parameter dispersion. The consequent reduction of read margins makes reliability more difficult to guarantee.

Each specific writing mechanism is associated to a set of reliability issues. *Tail bits* and *fast bits* are other counterpart issues of the tunneling based writing operations. Both have a direct impact on yield and read margins. Furthermore, the progressive increase of cell density creates new interdependent failure mechanisms (Gregori et al. 2003). A potential failure mode for tunneling based writing mechanisms is related to the *erratic bit* behaviour, which consists in a sudden and unpredictable change of the tunneling current and therefore of the final value of the cell state parameter. Erratic erase may induce random in-field failures during cycling thus threatening the reliability of the array.

The way the cells are interconnected together in a memory architecture can also affect the memory reliability. In fact, the state of a cell which has never been written, could, in principle, be changed by physical mechanisms triggered by the voltages applied during reading or programming of other cells sharing some signal lines with the considered cell. This undesired change of the state of the cell goes under the name of *disturb* and can occur during both programming and reading operations. Disturbs can depend on the specific writing/reading waveforms and, in general, are more critical for higher signal frequencies.

In this scenario the on-chip ECC approach seems to play an important role in order to ensure the adequate reliability as NV (Non Volatile) memories move toward higher density and faster data transfer (Benjauthrit et al. 1996; Campardo et al. 2000).

5.2 Reliability issues in floating gate memories

In this section, after introducing some basic definitions, the most important reliability issues of floating gate memories will be described considering first the single cell intrinsic effects and then the array dependent reliability issues.

5.2.1 Floating gate memory basic definitions

In principle a Flash memory cell can be thought as a traditional MOSFET transistor structure with the addition of a floating gate (Fig. 5.1).

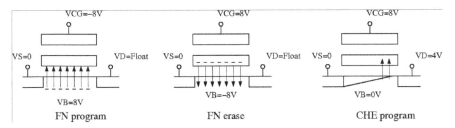

Fig. 5.1. Representation of the floating gate cell structure and examples of cell polarizations during writing using the two physical mechanisms: FN and CHE

The single cell is a four terminal device: control gate, source, drain and bulk. The state of the cell can be controlled by varying the number of electrons stored on the floating gate. The larger is the number of electrons on the floating gate the higher is the threshold voltage of the equivalent MOSFET. Figure 5.2 shows a representation of the IV cell characteristics in the two states: programmed (floating gate filled by electrons) and erased (no electrons on the floating gate). The cell threshold voltage can be defined as the control gate voltage at which the cell drains a predefined current. Figure 5.2 shows also the programmed (VTP) and erased (VTE) threshold voltages as well as the separation between the two levels in terms of threshold voltage distance (Chimenton and Olivo 2005). A correct cell level sensing requires adequate read margins, and therefore, threshold voltage distributions have to be as tight as possible. This goal is however becoming more and more difficult to achieve, especially if geometry scaling (which already induces a threshold voltage spread increase) is associated to a power supply voltage reduction. Things are even worse for ML architectures where level separation is further reduced.

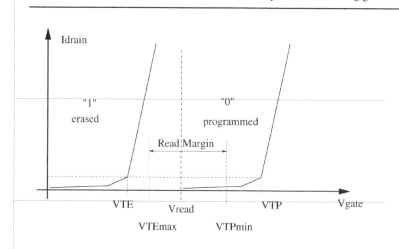

Fig. 5.2. Read margin definition in the IV cell characteristics (Chimenton and Olivo 2005)

In general, in a ML architecture, the number of digital states is 2^b where b is the number of bits per cell. From a physical point of view each level differs by the number of electrons stored in the floating gate.

There are two common ways to connect floating gate cells together in order to build an addressable array of cells: the NOR architecture and the NAND architecture.

In a NOR architecture, cells are connected in parallel to the bitlines as shown in Fig. 5.3 and programming is achieved via the Channel Hot Electron (CHE) physical mechanism by applying a high voltage (e.g. 8–10 V) to the control gate and 4–5 V to the drain (see Fig. 5.1). Source and bulk are grounded or held at a negative voltage, thus preventing reading errors caused by excessive bitline leakage. Erasing is achieved via FN (Fowler-Nordheim) tunneling (see Fig. 5.1). Drain is commonly left floating during erase. Erasing involves sectors or blocks of cells while programming involves bytes or groups of bytes/words. In NAND architectures, cells are connected in series to the bitline through two selecting transistors (see Fig. 5.3). Both programming and erasing are carried out using FN tunneling. CHE is used to inject electrons in the floating gate from the channel of the equivalent MOSFET (see Fig. 5.4). Near the drain, channel electrons can gain sufficient energy to jump over the silicon dioxide barrier and some of them can reach the floating gate. FN tunneling can be used for both injection or removal of electrons into/from the floating gate (see Fig. 5.4). Electrons can tunnel the silicon dioxide barrier due to the large electric fields. The physics behind the two mechanisms is not here reported, anyway it is important to note that both Flash basic writing operations require high voltages to be effective, thus potentially inducing several types of degradations and reliability issues.

Another important difference between the two architectures is that NAND is intrinsically much more compact than NOR, since no drain contact areas are necessary between the cells.

Although NAND architectures are slower than NOR, they are intrinsically more dense and thus more preferable in all data storage applications, whereas NOR is preferable in code storage application.

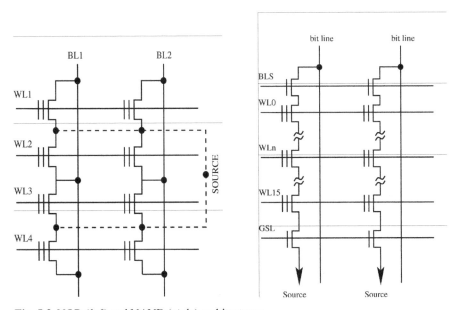

Fig. 5.3. NOR (*left*) and NAND (*right*) architectures

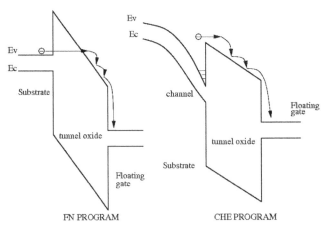

Fig. 5.4. Physical mechanisms used to inject-remove electrons to-from the floating gate: FN tunnelling (*left*); CHE (*right*)

As shown in Fig. 5.5 the number of electrons stored in the floating gate scales down with the litho node. Several reliability issues may steam from dealing with a small number of electrons per memory state. In particular, for a given data retention specification, lower oxide leakage currents have to be guaranteed and this means that the quality of tunnel oxides, which is already very high, has to be continuously improved.

The same figure shows also that reliability requirements are more critical for NAND since, for a given litho node, NAND requires a larger number of electrons compared to NOR. In fact, tunnel oxide degradation during cycling can be larger in NAND as FN tunneling, and therefore Stress Induced Leakage (SILC) (Olivo et al. 1988), occurs during both program and erase operation.

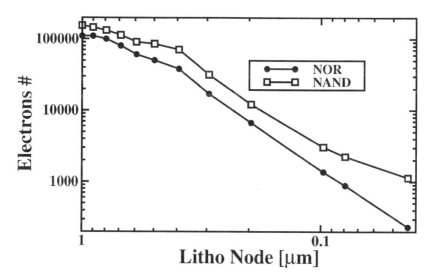

Fig. 5.5. Number of electrons stored in the floating gate as a function of the litho-graphic node (data from (Lai 2004))

5.2.2 Intrinsic oxide degradation

Fowler-Nordheim in Flash memories intrinsically causes oxide degradation. This is mainly due to the high fields which cause charge trapping inside the tunnel oxide (Park and Schroeder 1998). There are many experimental evidences of charge generation and trapping within the oxide. The typical gate voltage turnaround effect is well known: for a fixed tunneling current, a negative shift of the gate voltages indicates hole trapping while, after a certain amount of stress, the increase of the gate voltage indicates electron trapping. Accumulation of negative charge gives rise to the well known program/erase window closure (Modelli et al. 2004) which basically limits the maximum number of program/erase cycles (see Fig. 5.6).

Electron trapping in the oxide has been shown to be reduced by process improvements. For example, oxide nitridation (Fukuda et al. 1991) and interpoly process optimization (Ushiyama et al. 1995) can improve memory endurance.

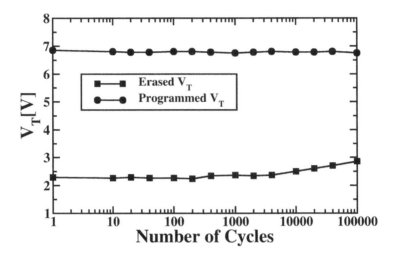

Fig. 5.6. Typical window-closure of NOR Flash cells (data from (Modelli et al. 2004))

As already recalled, other intrinsic degradations are due to holes generated during FN tunneling and then injected into the oxide. Hole injection causes erase instabilities like erratic erase, interface trap generations and therefore transconductance degradation (Yamada et al. 1993), increase of sub-threshold leakage (and therefore bitline leakage) and reduction of program efficiency.

Writing waveform optimization can help in removing the trapped charge. For example, it has been shown that window closure can be reduced by using low voltage erasing pulses able to remove the charge accumulated in the oxide (Lee et al. 1999).

Trapped charge reduction is also important for data retention. In fact, detrapping currents can significantly influence the threshold voltage of floating gate cells where information storage is based on a small number of electrons (Mielke et al. 2004). Detrapping mechanisms have to be understood in order to limit in-field data retention failures and the parallel use of ECC can in some cases be mandatory.

5.2.3 Parameter dispersion

Cell geometrical parameters need to be kept under strong process control for the memory to operate correctly. In fact the memory threshold voltage distribution width strongly depends on the dispersion/control of a finite set of critical geometrical/process parameters. Figure 5.7 shows both a programmed and an erased threshold voltage distributions of a 4 Mbit NOR array. Programming has been performed by using CHE whereas erasing has been performed by using FN tunneling. This figure represents the threshold voltage distribution in a Gaussian probability paper where an axis transformation allows Gaussian distributions to be represented by straight lines. A first observation is that CHE programming results in a narrower and Gaussian-like distribution compared to the FN-erased one. The FN-erasing distribution typically gives rise to a relatively large percentage of cells whose erasing behaviour significantly deviates from the normal-Gaussian part of the distribution. These anomalous cells are commonly referred to as *tail bits*. On the contrary, no anomalous cells of this kind can be seen in the CHE programmed distribution. It is also important to notice that both operations have been carried out using one single programming/erasing pulse, i.e. without performing any kind of *verify* procedure. This procedure aims at verifying that the threshold voltages of all the cells are lower or higher than a predefined value and can be applied to both FN or CHE operations. The erasing/programming step is thus repeated until the condition becomes true or a maximum number of pulses is reached (thus generating a program failure condition). Verify procedures have been introduced early in the Flash memory history in order to guarantee adequate reading margins and to reduce the risks related to the application of unnecessary long FN erasing pulses (see the overerase issue in the next sections) (Kynnet et al. 1989). The absence of the verify routine in both operations in the results of Fig. 5.7 is an important assumption for a comparison of the intrinsic effects of the two operations. Therefore, it can be noticed that the sensitivity to technological/process parameter dispersion is larger for a FN tunneling operation rather than for the CHE writing operation. Moreover, since CHE can be applied to individual bytes or words, the verify option performed during CHE allows the program distribution to be very tight, an extremely important condition in ML architectures.

It follows that read margins and possible related reliability issues can potentially be more serious for architectures using FN tunneling for both operations (i.e. NAND versus NOR).

Basically the reliability concerns about threshold voltage distribution widths can be: reduction of read margins; increase of total bitline leakage current during reading; increase of electrical stress and cell degradation due to post-erase algorithms (soft programming, post-erase repair and so on) introduced to compact the threshold voltage distribution.

Limits on reading margins are very important especially for ML and scaled technologies using low supply voltages (Oyama et al. 1992). For example, in a 3.3 V single bit technology, the target distribution width should be about 1 V in order to prevent bitline leakage. Such target widths are commonly achieved through the use of post erase algorithms.

The suppression of the bitline leakage current is important for both program and read operations and is of major importance in those technologies that do not make use of additional select transistors, e.g. NOR architectures. In general leakage currents are an issue for the boost circuits and charge pumps which provide the high voltages required during both writing and reading operations: a large bitline leakage current during programming can reduce the effective voltage at the drain contact in CHE operations because of the parasitic bitline voltage drop, thus slowing down the program operation and eventually causing a program failure; a large bitline leakage current during the sensing of the state of a programmed cell (cell A in Fig. 5.8) can give rise to a read failure (the programmed cell A erroneously appears in the erased state).

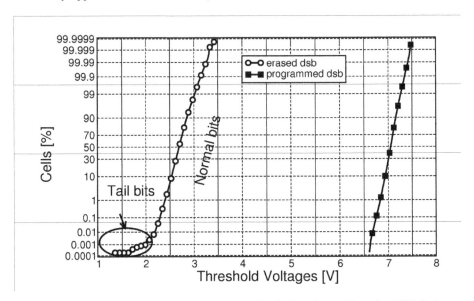

Fig. 5.7. *Right*: programmed threshold voltage distribution obtained by using CHE. *Left*: erased distribution obtained by using FN tunneling (Chimenton and Olivo 2005)

In order to reduce the bitline leakage current during reading or programming, all the unselected wordlines are kept at ground or are driven at a negative voltage (e.g. –1 V). In the presence of a large erased distribution width and/or of a bottom tail in the erased distribution, there is a significant probability to find some cells with very low thresholds that are difficult to turn-off. As a consequence, technology scaling requires to increase the absolute value of the unselected voltages which can induce cell degradation and stress. In particular, unselected cells may experience large voltages during programming as the drain (4 V) to gate (–2V) voltage could easily reach values of 6 V (see 'Stress voltage' in Fig. 5.8). Such large voltages can have serious consequences on device reliability. For example, they have been shown to induce data retention degradations due to hot hole injection (Chimenton et al. 2002).

In a NAND architecture, the presence of overerased cells, which behave as depleted MOSFET transistors, i.e. switches in the ON state, is not a particular issue since the reading operation requires that all unselected cells behave as switches in the ON state. On the other hand, over-programmed cells are equivalent to switches which are difficult to turn on. Therefore, over-programming has to be carefully avoided in NAND architectures.

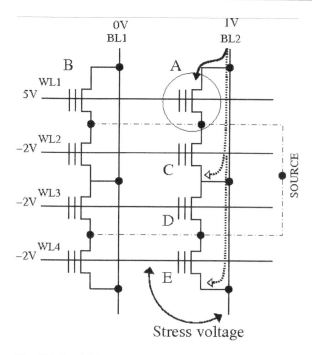

Fig. 5.8. Read failure caused by bitline leakage current (*dashed line*) while reading cell A in a NOR-architecture (Chimenton and Olivo 2005)

In order to guarantee safer read-margins and to suppress leakage (NOR) or to avoid over-programmed cells (NAND), it is common practice to perform a special operation which compacts the erased threshold voltage distribution. These algorithms may have different names, e.g. Post-Erase Repair, Automatic Program Disturb, Soft-Programming and so on. Some drawbacks of these schemes are: increase of the total erasing time; additional area for the circuits implementing the post-erase algorithm; increase of power/energy consumption; increase of complexity and introduction of new reliability issues.

A more and more intensive use of post-erase routine is predicted as technology scales down and threshold voltage distributions become larger and larger.

In fact, the total bitline leakage current depends on the erased threshold voltage distribution probability density f and on the cell subthreshold leakage current (I_{sub}):

$$I_{MAXBL} = N \int_{VTC}^{\infty} f(VT) I_{sub}(VT) dVT \qquad (5.1)$$

where N is the number of cells in the bitline and VTC is the critical voltage at which the bitline current equals the maximum bitline leakage current I_{MAXBL}, above which reading/programming failures may occur.

Therefore all cells having $VT<VTC$ need to be compacted by the post-erase algorithm.

Their number is given by

$$N_{ov} = N \int_{-\infty}^{VTC} f(VT)dVT \qquad (5.2)$$

and is an increasing function of VTC. Therefore VTC should be minimized in order to reduce the post-erase routine usage. Indeed Fig. 5.9 shows the impact of the erased threshold voltage standard deviation σ on the critical voltage VTC calculated assuming an analytical approximation for the subthreshold currents.

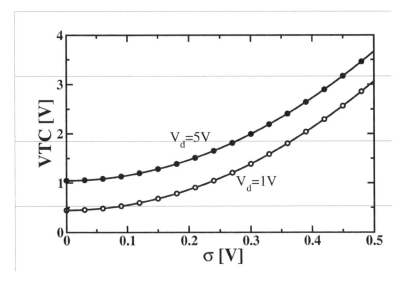

Fig. 5.9. Critical voltage VTC as a function of erased threshold voltage dispersion (Yoshikawa 1996)

As shown by the figure, the increase of threshold voltage width/dispersion induces a quadratic increase of the number of cells that need to be repaired. Note that Fig. 5.9 shows the net effect of the Gaussian part of the erased distribution only (see Fig. 5.7), thus neglecting the tail population which can give a significant contribution to the total number of cells that need to be repaired.

As a matter of fact, it has been observed that the threshold voltage dispersion increases with the reduction of the cell size/area (Yamada et al. 1991). Hence, reliability implications due to erase distribution width/dispersion are expected to become more and more important with each new generation and, as a consequence, a deeper study of new compacting algorithms is required.

The influence of parameter dispersion on the erased distribution width has been studied using analytical approaches as well as numerical simulations. The specific cell structure needs to be taken into account properly, since different critical geometrical parameters may be involved. For a given erasing scheme it is possible to find a few dominant parameters, whose process control should be maximized whereas different erasing schemes or voltages can significantly change the set of dominant parameters.

Also other physical mechanisms occurring during a writing operation can play a significant role in the reliability of Flash cells. Holes can be trapped in the tunnel oxide, thus causing transconductance degradation, data retention degradation and gate disturbs. On the other side, the hole injection mechanism is responsible for interface state generation and transconductance degradation (Verma and Mielke 1988; Haddad et al. 1989; Chen et al. 1995). Holes trapped above the drain junction can degrade the performance of the program operation (Chen et al. 1995). In specific operating voltage conditions, holes can be injected at the drain junction, due to the large lateral electric fields, giving rise to the so called hot hole injection (HHI). Scaling the cell size reduces the channel length and enhances the lateral electric field. Hence HHI could be a serious issue for scaled devices. During FN tunneling, holes can be injected in the oxide from the substrate-accumulation layer at the Si-SiO$_2$ interface (anode hole injection AHI). AHI occurring during FN is considered the main cause of erratic erase, i.e. the VTE instability occurring during cycling (Ong et al. 1993). Some of the injected hot holes can be trapped near the floating gate and induce a local reduction of the tunneling electron barrier. In this situation the cell erases faster than normal and, after erase, it exhibits relatively low VTE values. The recombination of trapped holes or hole-detrapping processes can restore the original, normal, erasing behaviour. Therefore VTE switches randomly from normal to fast values during cycling, as shown in Fig. 5.10, with both positive and negative shifts.

Non-uniform tunneling can be another cause of threshold voltage dispersion. This effect is due to the non perfect interface between floating gate and tunnel oxide. There are asperities locally enhancing the electric field: localized positive charge trapped in the oxide near the floating gate, polysilicon impurities diffusing in the oxide, general local oxide thinning. Non-uniform tunneling can cause VT dispersion in a way more evident in small tunneling areas.

In fact, let A be the tunneling area and σ the localized high-conductive spot area, then the relative increase of tunneling current due to the addition of one spot of extra-current can be expressed as

$$\frac{\Delta I}{I_{avg}} = \frac{\sigma \Delta J}{n \sigma \Delta J + A J_N} \tag{5.3}$$

where J_N is the normal tunnel current density associated to an ideal region with no extra-currents, ΔJ is the extra-current at an anomalous spot with section σ, and n is the average number of anomalous spots per cell. As shown by Eq. (5.3), a reduction of the tunneling area A increases the ratio of the tunneling current deviation versus average current. Figure 5.11 shows an experimental evidence of non-uniform tunneling causing *VTE* dispersion with the reduction of the tunneling area.

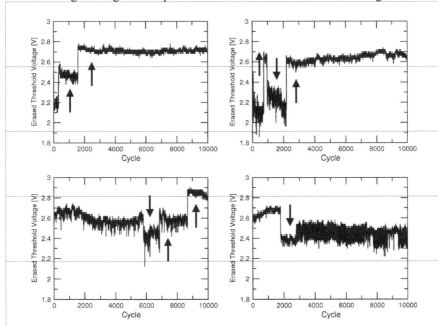

Fig. 5.10. Example of 4 erratic cells. The overall threshold voltage distribution width can be influenced by the erratic phenomenon (Chimenton et al. 2001)

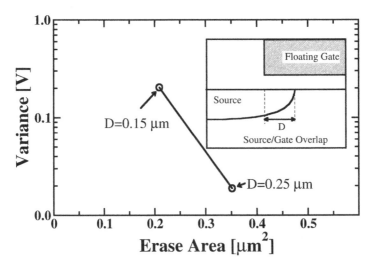

Fig. 5.11. Impact of tunneling area on *VTE* deviations (data from (Muramatsu et al. 1994))

Another cause of the threshold voltage distribution spread is due to contact misalignment. Due to the fact that cells closer to the bitline contact have a reduced erasing speed, two different distributions with different average values are present after erase (one for the odd rows and one for the even rows), thus causing a global increase in the *VTE* dispersion.

In conclusion, the particular choices of erasing conditions as well as the correct control of critical parameters are fundamental in controlling the erase distribution width. Non-uniform tunneling and charge trapping during both FN and CHE writing operations are becoming a dominant cause of threshold voltage deviations and reliability issues in scaled technologies.

5.2.4 Erratic bits

Erratic bits are cells that show an unstable and random *VTE* during cycling. Shifts larger than one Volt can easily be measured during program-erase cycles. Figure 5.10 shows the erased threshold voltage of four different cells during cycling. When an erratic bit switches to a fast state, it can undergo overerase, i.e. its final threshold voltage after the FN erase operation can be extremely low. In NOR-architectures this anomalous behaviour can directly lead to single bit failures due to the extra-bitline leakage current caused by the overerased cell. In NAND-architectures, since FN is applied for both program and erase, it can give rise to over-programmed cells which are difficult to turn-on, thus slowing down the reading operation of the string in which the cell is connected.

Erratic erase has been attributed to hole trapping/detrapping in the oxide (Ong et al. 1993; Chimenton et al. 2003) close to the floating gate (see Fig. 5.12) which is able to dramatically increase the tunneling current (Dunn et al. 1994) (see Fig. 5.13).

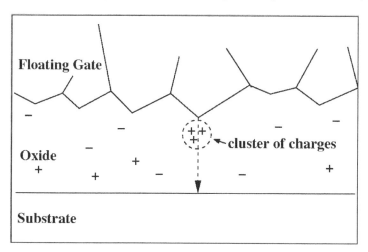

Fig. 5.12. The positive charge cluster model assumed to explain erratic erase (Chimenton et al. 2003)

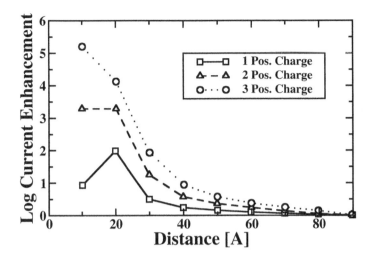

Fig. 5.13. Tunneling current enhancement due to one/two or three charge clusters as a function of its position in the oxide from the floating gate (data from (Dunn et al. 1994))

Several experimental evidences are in agreement with this hypothesis. For example, it has been shown that a 250°C bake or an Ultra-Violet exposure can turn some erratic bits into normal bits (Ong et al. 1993; Chimenton and Olivo 2003). However the origin of the positive charge cluster still remains debated. Initially, the positive charge has been attributed to holes generated by Band-to-Band Hot Hole Injection during source-side erase (San et al. 1995). Indeed erratic erase has been observed also during channel erase where the source to bulk voltage is too low for the hot hole generation to occur (Chimenton et al. 2003).

Another hypothesis is that impact ionization can occur in the silicon dioxide during FN operation. However, impact ionization is more likely to occur in oxides thicker than that used in today's Flash memories ranging from 7 to 10 nm.

Today's most likely hypothesis on the origin of the charge cluster is related to an Anode Hole Injection mechanism (Chimenton and Olivo 2002). As shown in Fig. 5.14, electrons tunneling from the floating gate generate hole-electron pairs in the substrate (anode). High energetic holes are then injected in the oxide where they can be trapped near the floating gate, thus creating the cluster of charges.

Many experimental results are in agreement with the AHI hypothesis: erratic erase has been observed by reversing the voltage polarity, i.e. by injecting electrons from the channel to the floating gate (Chimenton et al. 2003); a smaller number of erratic bits observed when operating with thin tunnel oxides (Chimenton and Olivo 2002); the increase of the number of erratic bits with larger erasing electric fields (Chimenton and Olivo 2004).

One of the most harmful features of the erratic behaviour is its capability to generate in-field reliability issues. Erratic bits can behave as normal bits and can therefore pass common screen tests. Nevertheless, analysis and experimental results show that all the cells of an array can potentially exhibit erratic behaviours in a long cycling experiment. Erratic behaviour is therefore a cell intrinsic behaviour which is not related to specific defects, and which cannot be completely eliminated.

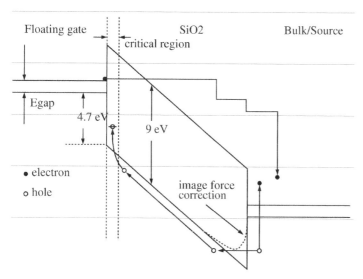

Fig. 5.14. The Anode Hole Injection mechanism generating the cluster of charges responsible for the erratic erase (Chimenton and Olivo 2005)

Erratic behaviour shows up as a secondary effect of the FN tunneling mechanism. Single bit over-erase or over-programming issues caused by erratic bits are commonly overcame by compacting routines (soft-program, drain disturb etc.) which aim at compacting the threshold distribution. Nevertheless, in an optimal design, it is difficult to properly take into account the actual occurrence of erratic behaviours and this can result in non optimal circuits with oversized area and power. Note also that post-repair algorithms can have an added impact on cell reliability as the cell degradation due to their use is still to be completely understood and assessed.

ECC can therefore be used as an additional strategy able to face erratic phenomena, whenever the use of compacting algorithms becomes less effective. In the case a normal bit turns into a fast bit, instead of carrying out the whole post-erase algorithm (which repairs all the cells of the bitline), it can be more convenient to simply correct via ECC the anomalous readings of the other cells on the same bitline. ECC convenience over compacting algorithms should take into account the statistical properties of the erratic phenomena. For example, experimental data shows that not all erratic bits behave in the same way: the log-log plot of Fig. 5.15 shows how the erratic event occurrence is distributed among erratic cells.

And not all erratic events will eventually cause a failure: a typical distribution of threshold voltage deviations is shown in Fig. 5.16. Since the cell switches from a normal to a fast condition and vice versa, a positive or a negative shift can result during an erratic event as shown by the symmetry of the distribution of Fig. 5.16 and large shifts are much less probable than smaller ones.

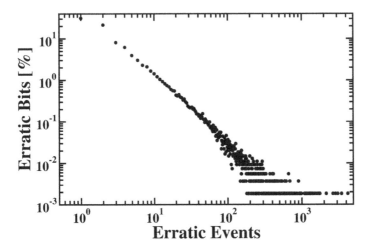

Fig. 5.15. Erratic event occurrence: only few cells exhibit a large number of erratic events (Chimenton et al. 2003)

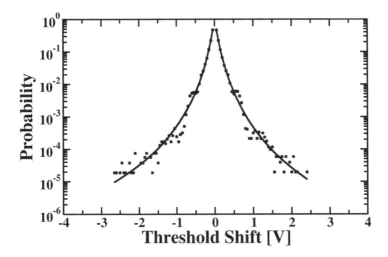

Fig. 5.16. Distribution of threshold voltage shifts due to erratic events (Chimenton and Olivo 2003)

It is important to notice that only negative shifts may induce single bit failures in NOR-architectures, while positive shifts may induce over-programming failures in NAND architectures. By taking into account these data it is possible to estimate the over erase/programming probability for a given number of cycles (see Fig. 5.17).

Beside post-repair algorithms and ECC, there are however also other methods that aim at reducing erratic phenomena which are based on writing waveform and process optimization. Writing waveform optimization can be of primary importance as it has been shown that different waveforms can have a different impact on erratic phenomena. For example, a Constant Charge Erasing Scheme (CCES) (Chimenton et al. 2002) allows to obtain a significant reduction of erratic bits with respect to a standard, constant voltage operation (Chimenton and Olivo 2003). Also the use of high-frequency pulsing can significantly improve the cell robustness to erratic phenomena (Chimenton and Olivo 2003). Process optimizations like oxide nitridation (Fukuda et al. 1991) or other process improvements suppressing hole generation and trapping can effectively be used to reduce the erratic phenomenon (Ushiyama et al. 1995; Muramatsu et al. 1994).

In general, due to the correlation between erratic bits and tail bits (as will be shown in the following), process improvements that reduce the tail of the erased distribution can also reduce erratic behaviours.

In conclusion, ECC is not the only solution against erratic phenomena. As it has been shown, there are many other possibilities that have to be investigated: the use of smaller tunnel oxide thickness; lower tunneling electric fields or high frequency writing pulsing (waveform optimization); other process improvements and post-repair algorithms. ECC is therefore an additional instrument that can be conveniently applied together with other solutions.

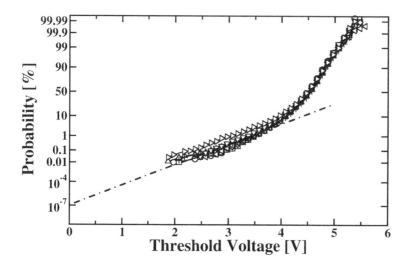

Fig. 5.17. Estimation of the over-erase probability by considering the distribution of threshold voltages after a negative erratic shift. The estimation of the over-programming probability can be performed in an analogous way (Chimenton and Olivo 2003)

5.2.5 Tail bits

Parameter dispersion and non-uniform tunneling are the main responsible for the threshold voltage scattering after tunneling based writing operations. In fact, Fig. 5.18 shows how a set (Set A) of 10,000 cells with the same average programmed threshold is scattered throughout the erased distribution after the FN writing operation. In particular, cells are scattered in a Gaussian way.

The figure shows also the initial distribution of tail bits in the programmed distribution: their distribution is gaussian with the same standard deviation of the other cells although a small negative shift reveals that tail bits are somehow less prone to be programmed (via CHE). Therefore, the same cell which exhibits an anomalous behaviour using a particular writing mechanism, does not necessarily show an anomalous behaviour also under other writing mechanisms.

It has been found that tail bits are strongly correlated with erratic behaviours. In fact, as shown in Fig. 5.19, tail bits are almost always also erratic bits.

Tail bits are related to non-uniform tunneling phenomena which can have various physical causes: irregular morphology of the injecting surface locally enhancing the electric field and/or positive charge clusters locally reducing the barrier height. Process improvements, for example the reduction of the poly-Si grain size in the floating gate, can reduce the tail of the erased distribution (Muramatsu et al. 1994; Nkansah and Hatalis 1999). Despite their anomalous behaviour, tail bits do not necessarily exhibit a significant charge loss and therefore they do not necessarily represent a concern for data retention (Scarpa et al. 2001).

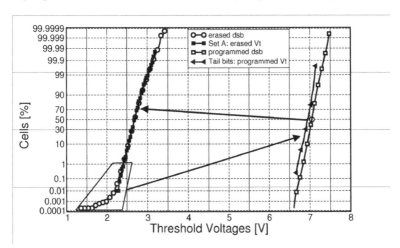

Fig. 5.18. Sub-distributions of tail bits before and after FN erase and normal bits (Chimenton and Olivo 2005)

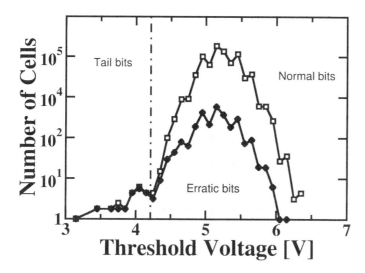

Fig. 5.19. Distribution of erratic bits (*grey*) compared with the overall cell population distribution (erratic bits + normal bits) (Chimenton et al. 2003)

5.2.6 Ultra fast bits

Ultra fast bits erase faster than both tail and normal cells in FN operations. Their threshold voltage can be several Volts below/above the tail of the erased/programmed distribution. Their percentage has to be very small as for the impact they can have on yield. Ultra fast bits are particular cells that do not necessarily show particular data retention problems, or anomalous values of transconductance, or of threshold voltage after both Ultra-Violet exposure and gate disturbs. Their position inside the cell matrix appears random and the erase characteristics of the neighbor cells can be normal. Both UV exposure and bake (up to 250°C) are not able to suppress or modify the ultra fast characteristic.

From these results, geometrical parameter deviations (as for example high coupling ratios or large tunneling area), as well as oxide-thinning or trapped charge, can be ruled out from the set of causes of ultra fast behaviour.

Ultra-fast behaviour can, indeed, be explained assuming the presence of localized high electric fields forming extra-tunneling current in small spots. As a consequence, fast bits can exhibit significant wear-out due to electron trapping (see Fig. 5.20) during cycling.

Nevertheless, other experimental results do not show a convergence to a normal behaviour. More experimental results are therefore needed in order to clarify their anomalous behaviour. Also, up to date cycling measurements have not adequately revealed a significant erratic behaviour in ultra fast bits.

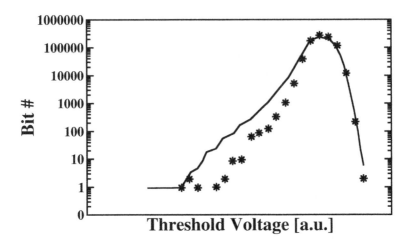

Fig. 5.20. Threshold voltage distribution before (*solid line*) and after (*symbols* *) cycling (data from (Cappelletti et al. 1994))

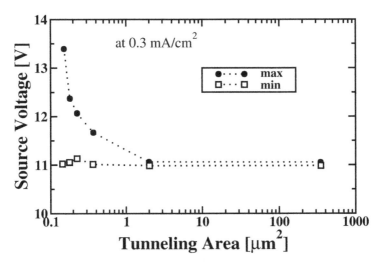

Fig. 5.21. Voltage spread as a function of the nominal tunneling area with constant tunneling current density (data from (Chimenton and Olivo 2004))

Hence, the origin of the anomalous tunneling spot has to be ascribed to a physical/morphological/geometrical cause rather than an electrical one. For example, surface roughness could be a consistent hypothesis.

In fact, the floating gate deposition step significantly affects the tunneling characteristics: doping dose, annealing temperature and doping species have shown to play important roles (Ushiyama et al. 1991). It has been observed that voltage deviations increase significantly with the reduction of the tunneling area (see Fig. 5.21).

This effect is consistent with the presence of localized high-conductive tunneling spots. In particular SEM and AFM images clearly showed the presence of oxide ridges at the floating gate/tunnel oxide interface (see Fig. 5.22).

Oxide ridges are assumed to be caused by phosphorous segregated in grain boundaries and diffused in the oxide, thus forming phosphorous rich oxide. The use of arsenic instead of phosphorous, as well as the reduction of annealing temperature/time and doping dose can improve the interface quality (Ushiyama et al. 1991) and reduce voltage deviations.

In conclusion ultra-fast bits seem to be related to extrinsic causes like floating gate-to-oxide spikes, whiskers or oxide ridges that locally enhance the electric field. Therefore, area reduction due to device scaling, can greatly amplify the threshold voltage deviations due to localized non-uniform tunneling. A careful process optimization has to reduce to the minimum the number of ultra fast bits.

Ultra fast bits can be easily identified during Electrical Wafer Sorting (EWS) due to their "permanent" anomalous properties (in contrast to erratic bits) and therefore, in principle, they can be eliminated through the use of redundancy. Indeed, there could be some cases in which some cells exhibit a behaviour which is intermediate between ultra-fast and normal. In these cases, and where redundancy has already been used, post-erase algorithms can try to repair these bits. In this operation, cell degradation may occur, particularly when the cell largely deviates from a normal behaviour. As a consequence, in-field errors can occur after a certain number of repairing operations. Hence, the support of ECC can provide a further safe operating margin.

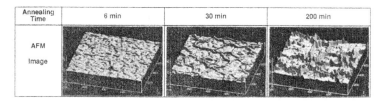

Fig. 5.22. Oxide ridges present at the floating gate to oxide interface and assumed to be responsible for ultra fast FN behaviour (Ushiyama et al. 1991)

5.2.7 Disturbs

Reading and writing disturbs are among the main reliability issues of scaled floating gate memories.

In general, disturbs are the result of the interconnection of the cells in the array which is therefore strongly architecture dependent. Consider for example the 9×12

array of Fig. 5.23 in which the cell at row 5 and column 7 is being programmed or read. During the operation, all cells sharing the same wordline (marked with 'W') experience the same voltages at their terminals. In the same way, another voltage configuration can be observed at the terminals of unselected cells of type 'U' and at those of type "B" (sharing the bitline). These parasitic voltage configurations can trigger disturbs in unselected cells and can depend on the voltages used during the required operation. For example, cells in the 'W' voltage configuration can experience the same control gate voltage of the cell being programmed (cell 'S'). If this voltage is too high, a gate disturb can occur. The other well known disturb is the so-called drain disturb which occurs in cells type 'B', i.e. cells sharing the same bitline. Nevertheless, many other configuration can occur when disturbs are supposed to be triggered not only by the considered cell voltage configuration, but also by the voltage configuration of neighbor cells (Chimenton and Olivo 2007). In addition, the final disturb can strongly depend also on the state of the considered cell, whereas neighbor cell state influence is in general lower, although, in more scaled technologies, it can soon start to show up significantly.

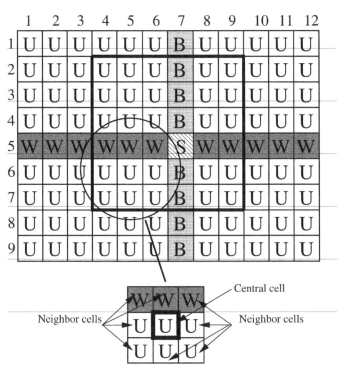

Fig. 5.23. Voltage configurations of a 9×12 matrix of cells during a program/read operation performed on cell (5,7) (Chimenton and Olivo 2007)

In general writing disturbs can have more serious effects than read disturbs, because of the use of higher operating voltages. Nevertheless, the total number of reading operations may be extremely high in some specific cell locations and they may give rise to an undesired threshold shift, especially for cells in a specific state which have been subjected to cycling-induced degradation. For example, in source-side erased NOR-Flash, it has been observed that cells in the erased state may appear programmed after a read disturb, because of channel electrons tunneling towards the floating gate through a lower oxide barrier. The cause of the oxide barrier lowering has been ascribed to holes trapped in the oxide near the substrate interface during common program/erase cycling (Brand et al. 1993). Also electron trapping has been found as physical cause for oxide degradation inducing read disturbs and also in this case program/erase cycling has found to weaken the read disturb immunity. Oxide nitridation has therefore been proposed as process improvement taking care of the problem (Kato et al. 1994). In ML architecture disturbs are made more severe by the longer programming time required for ML writing operations.

Since all possible disturb configurations cannot be tested for each specific wafer during common Electrical Wafer Sorting (EWS) at a production level, also in this case, ECC can be the only viable solution to take care of possible disturbs occurring during memory lifetime that can depend on the occurrence of particular conditions: state 0/1 of the cells in the array; values of voltage applied to the cells; presence of defects; level of degradation; process parameters at their corner cases.

5.3 Conclusions

The continuous increase of memory density poses several challenges to device reliability engineers. Failure analysis and physics of failure, which are the main instruments to be used for a successful achievement of a permanent reliability improvement, need an increasing effort in terms of economical investment and are both time and human resources demanding. It is becoming difficult to follow this trend on the basis of an ever increasing rate of technology scaling, roadmap and milestones requirements. In this scenario, particularly in the contest of the semiconductor NVM technology, the usage of ECC combined together with all other techniques described in this chapter can be the only viable solution for providing a good trade-off between performance and reliability, thus guaranteeing the necessary time-to-market of the final product.

Bibliography

A. Aritome, R. Shirota, G. Hemink, T. Endoh, F. Masoka, "Reliability Issues of Flash Memory Cells", *Proc. IEEE*, Vol. 81, pp. 776–788, May 1993.

B. Benjauthrit, L. Coday, M. Trcka, "An Overview of Error Control Codes for Data Storage" in *Proc. IEEE Int. Non-Volatile Memory Technology Conf.*, 1996, pp. 120–126.

A. Brand, K. Wu, S. Pan, D. Chin, "Novel Read Disturb Failure Mechanism Induced by FLASH Cycling", in *International Reliability Physics Symposium* pp. 127–132, 1993.

W. D. Brown, J. E. Brewer, Eds., Nonvolatile Semiconductor Memory Technology, *New York: IEEE Press,* 1998.

G. Campardo et. al., "40-mm^2 3-V-Only 50 MHz 64-Mb 2-b/Cell CHE NOR Flash Memory", *IEEE Journal of Solid State Circuits*, Vol. 35, No. 11, pp. 1655–1667, November 2000.

P. Cappelletti, R. Bez, D. Cantarelli, L. Fratin, "Failure Mechanisms of Flash Cell in Program/Erase Cycling", in *IEDM Tech. Dig.,* pp. 291–294, 1994.

P. Cappelletti, C. Golla, P. Olivo, E. Zanoni, Eds., "Flash Memories", *Boston, MA: Kluwer*, 1999, Ch. 5.

J. Chen, J. Hsu, S. Luan, Y. Tang, D. Liu, S. Haddad, C. Chang, S. Longcor, J. Lien, "Short Channel Enhanced Degradation During Discharge of Flash EEPROM Memory Cell", in *IEDM Tech. Dig.*, pp. 331–334, 1995.

A. Chimenton, P. Olivo, "Impact of Tunnel Oxide Thickness on Erratic Erase in Flash Memories", in *Proc. of the European Solid-State Device Conference*, pp. 363–366, 2002.

A. Chimenton, P. Olivo, "Erratic Erase In Flash Memories (Part II): Dependence on Operating Conditions", *IEEE Transaction on Electron Devices*, Vol. 50, No. 4, pp. 1015–1021, 2003.

A. Chimenton, P. Olivo, "Erratic Erase in Flash Memories (Part I): Basic Experimental and Statistical Characterization", in *IEEE Transaction on Electron Devices*, Vol. 50, No. 4, pp. 1009–1014, 2003.

A. Chimenton, P. Olivo, "Reliability of Flash Memory Erasing Operation under High Tunneling Electric Field", in *Proc. IRPS*, pp. 216–221, 2004.

A. Chimenton, P. Olivo, "Reliability of Erasing Operation in NOR-Flash Memories", introductory invited paper in *Microelectronics Reliability*, Vol. 45, No. 7–8, July–August, pp. 1094–1108, 2005.

A. Chimenton, P. Olivo, "Fast Identification of Critical Electrical Disturbs in Non-Volatile Memories", to appear in *IEEE Transaction on Electron Devices,* September 2007.

A. Chimenton, P. Pellati, P. Olivo, "Analysis of Erratic Bits in Flash Memories", *IEEE Transactions on Devices and Materials Reliability*, Vol. 1, pp. 179–184, December 2001.

A. Chimenton, P. Pellati, P. Olivo, "Constant Charge Erasing Scheme for Flash Memories", *IEEE Transactions on Electron Devices*, Vol. 49, pp. 613–618, April 2002.

A. Chimenton, P. Pellati, P. Olivo, "Overerase Phenomena: An Insight Into Flash Memory Reliability, " in *Proc. IEEE*, Vol. 91, pp. 617–626, April 2003.

A. Chimenton, P. Pellati, P. Olivo, "Erratic Bits in Flash Memories under Fowler-Nordheim Programming", *Japanese Journal of Applied Physics*, Vol. 42, No. 4B, pp. 2041–2043, 2003.

A. Chimenton, A. S. Spinelli, D. Ielmini, A. Lacaita, A. Visconti, P. Olivo, "Drain-accelerated Degradation of Tunnel Oxides in Flash Memories", in *IEDM Tech. Dig.*, pp. 167–170, 2002.

C. Dunn, C. Kaya, T. Lewis, T. Strauss, J. Schreck, P. Hefley, M. Middendorf, T. San, "Flash EPROM Disturb Mechanisms", in *Proc. IRPS*, pp. 299–308, 1994.

H. Fukuda, M. Yasuda, T. Iwabuchi, S. Ohno, "Novel N_2O-Oxynitridation Technology for Forming Highly Reliable EEPROM Tunnel Oxides Films", *IEEE Electron Device Letters*, Vol. 12, No. 11, pp. 587–589, 1991.

S. Gregori, A. Cabrini, O. Khouri, G. Torelli, "On-Chip Error Correcting Techniques for New-Generation Flash Memories" *Proc. IEEE*, Vol. 91, No. 4, pp. 602–616, April 2003.

S. Haddad, C. Chang, B. Swaminathan, J. Lien, "Degradations due to Hole Trapping in Flash Memory Cells", *IEEE Electron Device Letters*, Vol. 10, No. 3, pp. 117–119, 1989.

M. Kato, N. Miyiamoto, H. Kume, A. Satoh, T. Adachi, M. Ushiyama, K. Rimura, "Read-Disturb Degradation Mechanism due to Electron Trapping in the Tunnel Oxide for Low-Voltage Flash Memories", in *International Electron Device Meeting*, pp. 45–48, 1994.

K. Kim, J. Choi, "Future Outlook of NAND Flash Technology for 40 nm Node and Beyond" in *IEEE 21st Non-volatile Semiconductor Memory Workshop*, pp. 9–11, 2006.

V. N. Kynnet, M. L. Fandrich, J. Anderson, P. Dix, O. Jungroth, J. A. Kreifels, R. A. Lodenquai, B. Vajdic, S. Wells, M. D. Winston, L. Yang, "A 90-ns One-Million Erase/Program Cycle 1-Mbit Flash Memory" *IEEE Journal of Solid-State Circuits*, pp. 1259–1263, 1989.

S. Lai, "Continuing Moor's Law Cost Reduction In Non-volatile Semiconductor Memories" in *European Workshop on Innovative Mass Storage Technologies*, 2004.

J. H. Lee, K. R. Peng, J. R. Shih, S. H. Chen, J. K. Yeh, H. D. Su, M. C. Ho, D. S. Kuo, B. K. Liew, Jack Y. C. Sun, "Using Erase Self-Detrapped Effect To Eliminate the Flash Cell Program/Erase Cycling Vth Window Close", in *Proc. IRPS*, pp. 24–29, 1999.

N. Mielke, H. Belgal, I. Kalastirsky, P. Kalavade, A. Kurtz, Q. Meng, N. Righos, J. Wu, "Flash EEPROM Threshold Instabilities due to Charge Trapping During Program/Erase Cycling", *IEEE Transaction on Devices and Materials Reliability*, Vol. 4, No. 3, pp. 335–344, 2004.

A. Modelli, A. Visconti, R. Bez, "Advanced Flash Memory Reliability", in *IEEE International Conference on Integrated Circuit Design and Technology*, pp. 211–218, 2004.

S. Muramatsu, T. Kubota, N. Nishio, H. Shirai, M. Matsuo, N. Kodama, M. Horikawa, S. Saito, K. Arai, T. Okazawa, "The Solution of Over-Erase Problem Controlling Poly-Si Grain Size -Modified Scaling Principles for FLASH Memory-", in *IEDM Tech. Dig.*, pp. 847–850, 1994.

T. Nakayama, Y. Miyawaki, K. Kobayashi, Y. Terada, H. Arima, T. Matsukawa, T. Yoshihara, "A 5-V Only One-Transistor 256 K EEPROM with Page-Mode Erase", *IEEE Journal of Solid-State Circuits*, Vol. 24, pp. 911–915, August 1989.

F. D. Nkansah, M. Hatalis, "Effects of Flash EEPROM Floating Gate Morphology on Electrical Behavior of Fast Programming Bits", *IEEE Transaction on Electron Devices*, Vol. 46, No. 7, pp. 1355–1362, 1999.

P. Olivo, T. Nguyen, B. Riccò, "High Field Induced Degradation in Ultra-Thin SiO2 Films" in IEEE Transactions on Electron Devices, Vol. 35, pp. 2259–2265, 1988.

T. C. Ong, A. Fazio, N. Mielke, S. Pan, N. Righos, G. Atwood, S. Lai, "Erratic Erase in ETOX™ Flash Memory Array", in *VLSI Symp. on Tech.*, pp. 83–84, 1993.

K. Oyama, H. Shirai, N. Kodama, K. Kanamori, K. Saitoh, Y.S. Hisamune, T.Okazawa, "A Novel Erasing Technology for 3.3 V Flash Memory with 64 Mb Capacity and Beyond", in *IEDM Tech. Dig.*, pp. 607–610, 1992.

J.-H. Park, S.-H. Hur, J.-H. Leex, J.-T. Park; J.-S. Sel, J.-W. Kim, S.-B. Song, J.-Y. Lee, J.-H. Lee, S.-J, Son, Y.-S. Kim, M.-C. Park, S.-J. Chai, J.-D. Choi, U.-I. Chung, J.-T. Moon, K.-T. Kim, K. Kim, B.-I. Ryu, "8 Gb MLC (Multi-Level Cell) NAND Flash Memory Using 63 nm Process Technology" in *IEEE International Electron Devices Meeting*, pp. 873–876, 2004.

Y. B. Park, D. K. Schroeder, "Degradation of Thin Tunnel Gate Oxide Under Constant Fowler-Nordheim Current Stress for a Flash EEPROM", *IEEE Transaction on Electron Devices*, Vol. 45, pp. 1361–1368, June 1998.

P. Pavan, R. Bez, P. Olivo, E. Zanoni, "Flash Memory Cells-An Overview", *Proc. IEEE*, Vol. 85, pp. 1248–1271, August 1997.

T. K. San, C. Kaya, T. P. Ma, "Effects of Erase Source Bias on Flash EPROM Reliability", *IEEE Transaction on Electron Devices*, Vol. 42, No. 1, pp. 150–159, 1995.

A. Scarpa, G. Tao, J. Dijkstra, F. G. Kuper, "Tail Bit Implications in Advanced 2 Transistors-Flash Memory Device Reliability", in *Microelectronic Engineering* , Vol. 59, No. 1–4, pp. 183–188, 2001.

T. Tanzawa, T. Tanaka, K. Takekuchi, R. Shirota, S. Aritome, H. Watanabe, G. Hemink, K. Shimizu, S. Sato, Y. Takekuchi, K. Ohuchi, "A Compact On-Chip ECC for Low Cost Flash Memories", *IEEE Journal of Solid-State Circuits*, Vol. 32, pp. 662–669, May 1997.

M. Ushiyama, H. Miura, H. Yashima, T. Adachi, T. Nishimoto, K. Komori, Y. Kamigaki, M. Kato, H. Kume, Y. Ohji, "Improving Program/Erase Endurance by Controlling the Inter-poly Process in Flash Memory", in *Proc. IRPS*, pp. 18–23, 1995.

M. Ushiyama, Y. Ohji, T. Nishimoto, K. Komori, H. Murakoshi, H. Kume, S. Tachi, "Two Dimensionally Inhomogeneous Structure at Gate Electrode/Gate Insulator Interface Causing Fowler-Nordheim Current Deviation in Nonvolatile Memory", in *Proc. IRPS*, pp. 331–336, 1991.

R. Vancu, L. Chen, R. L. Wan, T. Nguyen, C.-Y. Yang, W.-P.Lai, K.-F.Tang, A. Minhea, A. Renninger, G. Smarandoiu, "A 35 ns 256 k CMOS EEPROM with Error Correcting Circuitry", in *IEEE Int. Solid-State Circuits Conf. Dig. Tech. Papers*, 1990, pp. 64–65.

G. Verma, N. Mielke, "Reliability Performance of ETOX Based Flash Memories", in *Proc. IRPS*, pp. 158–166, 1988.

S. Yamada, Y. Hiura, T. Yamane, K. Amemiya, Y. Ohshima, K. Yoshikawa, "Degradation Mechanism of Flash EEPROM programming after program/erase cycles", in *IEDM Tech. Dig.*, pp. 23–26, 1993.

S. Yamada, T. Suzuki, E. Obi, M. Oshikiri, K. Naruke, M. Wada, "A Self-Convergence Erasing Scheme for Simple Stacked Gate Flash EEPROM", in *IEDM Tech. Dig.*, pp. 307–310, 1991.

K. Yoshikawa, "Impact of Cell Threshold Voltage Distribution in the Array of Flash Memories on Scaled and Multilevel Flash Cell Design", in *Symposium on VLSI Technology Digest of Technical Papers*, pp. 241–242, 1996.

6 Hardware implementation of Galois field operators

6.1 Gray map

The alphabet used in coding theory can be seen as a set of equally spaced points on a circumference. Generally, in communication channels, noise does not make all the errors equally probable, that is, it is much easier that a transmitted symbol is erroneously received as a near symbol rather than a far one. For example, it is much easier that on sending 4 we receive 3 or 5 rather than 1 or 10. With this hypothesis the Hamming distance (Chap. 1, Definition 1.5.7) is not the most suitable metric to measure errors.

Definition 6.1.1 Considered an alphabet in $GF(m)$, the *Lee weight* w_L of an integer i is defined by

$$w_L(i) = \min\{i, m - i\} \tag{6.1}$$

Definition 6.1.2 The *Lee distance* $d_L(x,y)$ is given by $w_L(x-y)$.

Suppose we are working in $GF(4)$ with an alphabet composed of 0, 1, 2 and 3. Using the metric defined by the Lee distance, we note that for 0, 1 and 2 the weights are identical to the elements themselves, while this is not true for 3. In fact, 0 has weight 0, 1 has weight 1, 2 has weight 2, while 3 has weight 1.

In the case of multilevel memories, where the cell may be in one of the 2^m possible states, the first step in designing an ECC is encoding the alphabet of 2^m symbols through a binary alphabet.

The basic idea of Gray encoding is to map blocks of symbols of codewords in blocks of channel symbols, so that the most probable errors in the channel symbols cause a minimum number of error symbols in the codewords. In other words, we look for a natural map between $GF(2^m)$ and $GF(2)$ so as to map the metrics constructed in the two fields, that is the Lee metrics into the Hamming metrics.

For example, we want to construct a Gray map between $GF(4)$ and $GF(2)$. We introduce 3 functions from $GF(4)$ towards $GF(2)$ as described in Table 6.1.

Note that i written in the binary system is $(\alpha(i),\beta(i))$, besides $\varphi(i)=\alpha(i)+\beta(i)$, where the sum is intended to be a binary sum, that is an exclusive-OR.

Table 6.1. Functions needed to map $GF(4)$ into $GF(2)$

$i \in GF(4)$	$\alpha(i)$	$\beta(i)$	$\varphi(i)$	$w_L(i)$
0	0	0	0	0
1	1	0	1	1
2	0	1	1	2
3	1	1	0	1

Definition 6.1.3   The *Gray map* is defined as $\theta(i)=(\beta(i),\varphi(i))$. The code $C'= \theta(C)$ is called *binary image* of C.

Note that generally C' is not linear since the Gray map is not linear. The crucial point is that the Gray map preserves the distance, that is, the Lee distance of two codewords in C is equal to the Hamming distance of their images.

Accordingly, the 4 states of a ML memory are respectively encoded as 11 10 00 01. Note that, using the decimal coding, the states would not have been encoded in the same way.

Definition 6.1.4 It is defined as *decimal coding* between $GF(2^m)$ and $GF(2)$ the map that transcribes each decimal symbol of the alphabet $GF(2^m)$ in binary form.

In the case of decimal encoding of $GF(4)$, the states would have been encoded as 11 10 01 00.

With decimal encoding there is no continuity when shifting from one state to another (observe that the central transition involves an error of 2 bits). On the contrary, with the Gray encoding the transition from a state to another consecutive one involves an error of only one bit. Applying an error correction code with both encodings, we see that a code corrector of only one bit corrects all the transitions between adjacent states if a Gray encoding is used, while this is not true in the case of decimal encoding for which a code corrector of 2 bits is required.

In conclusion, the construction of the Gray map is equivalent to encode the states, so that the transition between adjacent states involves the change of only one bit.

6.2 Adders

As explained in Chap. 2, cyclic codes exist in Galois fields. To explain their implementation it is therefore fundamental to describe the hardware implementation of the Galois field operations.

It is easier to implement the arithmetic operations in the Galois field than the usual arithmetic operations, as they do not have any carries.

In order to describe them we need to keep in mind all the possible representations of the field elements, that is, the polynomial representation, the vector representation and the exponential representation (Chap. 2, Table 2.5).

In the case of the sum in $GF(q^m)$ the most useful representations are the vector and the polynomial ones. Adding two elements of the Galois field is equivalent to adding two polynomials, that is:

$$\left(a_0 + a_1 q + \ldots + a_{m-1}q^{m-1}\right) + \left(b_0 + b_1 q + \ldots + b_{m-1}q^{m-1}\right)$$
$$= c_0 + c_1 q + \ldots + c_{m-1}q^{m-1}, \tag{6.2}$$

where

$$c_i = a_i + b_i \qquad for \quad i = 0, \ldots, m-1 \tag{6.3}$$

If we consider only Galois fields of the type $GF(2^m)$ we have

$$\begin{cases} c_i = 0 & \text{if } a_i = b_i \\ c_i = 1 & \text{if } a_i \neq b_i \end{cases} \tag{6.4}$$

Thus, two Galois field elements are added either by modulo-two addition of their coefficients or, in binary form, producing the bit-by-bit exclusive-OR function of the two binary numbers.

The operation performed via software is equivalent to the binary sum of vector representations. The sum is implemented via hardware by the circuit shown in Fig. 6.1.

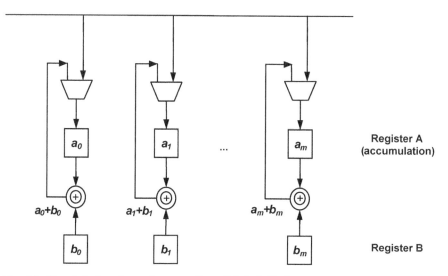

Fig. 6.1. Circuit that performs the sum of two Galois field elements

At the beginning the vector representation of the two elements to be added is loaded into the registers A and B. Their vector sum then appears at the input of register A. When the clock is pulsed, the sum is loaded into register A (register A is used as an accumulator).

In Appendix B (Sect. B.5.1) a routine implementing the sum is described.

6.3 Constant multipliers

The multiplication is a much more complicated operation than the addition. By multiplying two polynomials of degree $m-1$ we get a polynomial of degree $2m-2$ which is an invalid element of the Galois field.

For this reason, said $p(x)$ the field generator polynomial, the multiplication is defined as the *product modulo p(x)*. The *product modulo p(x)* is obtained by dividing the product of the polynomials by $p(x)$ and then keeping the remainder. Defined in this way, it is certain that the result is always of degree smaller than or at most equal to $m-1$ and therefore a valid element of the field.

First of all, consider the multiplication by a fixed element of the Galois field. Suppose we are working in $GF(2^4)$, whose primitive polynomial is $p(x)=1+x+x^4$. We want to multiply β in $GF(2^4)$ by the fixed field element α^3.

To perform the multiplication in an efficient way, it is convenient to use the exponential representation. In fact, β will correspond to the exponential representation α^i and the multiplication is easily performed by adding the exponents, so the result of the product will be:

$$\delta = \beta \cdot \alpha^3 = \alpha^i \cdot \alpha^3 = \alpha^{3+i} \tag{6.5}$$

To perform this operation via software we need two tables. A table is necessary to have the exponential representation of β as α^i. Once the exponents are added, it is necessary to have the result as vector representation, so we need the other table, which lets us easily go from the exponential representation toward the vector representation. Substantially, the operation performed via software requires the storage of two tables of 2^m-1 elements.

In case we want to perform this operation via hardware, we need to implement the circuit performing the multiplication.

Since the hardware implementation of the adder requires the vector representation, we use this representation also for the multiplier.

Still considering the example described above, the element β is represented as

$$\beta = b_0 + b_1\alpha + b_2\alpha^2 + b_3\alpha \tag{6.6}$$

and the multiplication will be performed as

$$\alpha^3\beta = b_0\alpha^3 + b_1\alpha^4 + b_2\alpha^5 + b_3\alpha^6 \tag{6.7}$$

Remembering that the primitive element satisfies the minimum polynomial of the field we get the equality

$$\alpha^4 = \alpha + 1 \tag{6.8}$$

Substituting Eq. (6.8) in Eq. (6.7) we get:

$$\alpha^3 \beta = b_1 + (b_1 + b_2)\cdot\alpha + (b_2 + b_3)\cdot\alpha^2 + (b_0 + b_3)\cdot\alpha^3 \tag{6.9}$$

Taking into account this expression, it is possible to build the circuit represented in Fig. 6.2 which is able to multiply any element β in $GF(2^4)$ by α^3. To perform the multiplication, we first load the vector representation (b_0, b_1, b_2, b_3) of β into the register b_i. Then we pulse the clock. The new contents of the registers will be the vector representation of $\alpha_3\beta$.

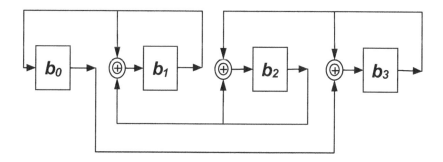

Fig. 6.2. Circuit that performs the product by α^3 of a field element in $GF(16)$

In general, every multiplication rule can be represented by a dedicated table. As an example, we will build the table in $GF(2^4)$ of the multiplication by the constant α^{12} represented in polynomial form (Fig. 6.3) as $\alpha^3 + \alpha^2 + \alpha + 1$.

The input bits contributing to a particular output bit are identified by the sum at the end of each column. Since the result can have degree 3 at most, the coefficients of degree 4, 5 and 6 must be written in the field, according to the field rule dictated by Eq. (6.8).

As the additions are modulo 2, these are implemented with exclusive-OR gates, as shown in Fig. 6.4.

Alternatively, each multiplier may be implemented as a look-up table with 2^m inputs which contains the result of the multiplication of each element by the constant. Obviously, this way becomes prohibitive as m increases.

	α^6	α^5	α^4	α^3	α^2	α^1	α^0
α^3	b_3	b_2	b_1	b_0	0	0	0
α^2		b_3	b_2	b_1	b_0	0	0
α^1			b_3	b_2	b_1	b_0	0
α^0				b_3	b_2	b_1	b_0
	b_3	b_2+b_3	$b_1+b_2+b_3$	0	0	$b_1+b_2+b_3$	$b_1+b_2+b_3$
				0	b_2+b_3	b_2+b_3	0
				b_3	b_3	0	0
				$b_0+b_1+b_2$	b_0+b_1	b_0	$b_0+b_1+b_2+b_3$

Fig. 6.3. Rule for the multiplication by α^{12} in $GF(16)$

A particular case concerns the multiplication in $GF(2^m)$ of an element β, represented by a polynomial of degree i, by x^j where $j<m-i$. The multiplication does not need to be performed modulo the field generator polynomial, since, as indicated by Eq. (6.10) the resultant element has degree equal to $m-1$ at most.

$$x^j \beta = x^j \left(a_0 + a_1 x + \ldots + a_i x^i \right) = a_0 x^j + a_1 x^{j+1} + \ldots + a_i x^{i+j} \qquad (6.10)$$

Thus, the multiplication is reduced to a shift of the element β to the right of j positions.

Hereafter, we will see the construction of a generic multiplier, used to multiply any two field elements.

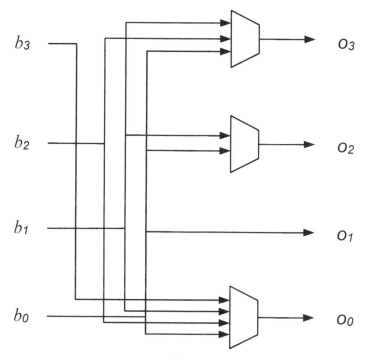

Fig. 6.4. Constant Multiplier by α^{12}

6.4 Full multipliers

When two generic elements of a Galois field $GF(q^m)$ are multiplied, the same considerations already made for the constant multiplier apply. In fact, also in this case, the multiplication is performed *modulo p(x)* as said above.

In order to implement this operation via software, it is useful to use the exponential representation. Similarly to the multiplication by constant, the method used consists in looking into a table for the representations of the two elements β and γ as powers of α and then adding their exponents. Finally, the result in exponential form is sought in the table again to have its vector representation.

$$\beta \cdot \gamma = \alpha^i \cdot \alpha^j = \alpha^{i+j} = \sigma \qquad (6.11)$$

The hardware required to develop this operation is more complicated compared to the multiplication by constant. Always considering the case of $GF(2^4)$, two elements β and γ are expressed in polynomial form as

$$\beta = b_0 + b_1\alpha + b_2\alpha^2 + b_3\alpha^3 \qquad (6.12)$$

$$\gamma = c_0 + c_1\alpha + c_2\alpha^2 + c_3\alpha^3 \tag{6.13}$$

The product $\beta\gamma$ can be expressed in the form described by Eq. (6.14), called Horner rule:

$$\beta \cdot \gamma = \left(\left(\left(c_3\beta\right)\cdot\alpha + c_2\beta\right)\cdot\alpha + c_1\beta\right)\cdot\alpha + c_0\beta \tag{6.14}$$

The circuit for the implementation of this equation, represented in Fig. 6.5, is composed of two blocks. The first one represents the multiplication by constant α, while the second one performs the other operations indicated in Eq. (6.14).

At the beginning the feedback shift registers of the circuit are empty while (b_0, b_1, b_2, b_3) and (c_0, c_1, c_2, c_3), i.e. the vector representations of β and γ, are loaded into the registers B and C respectively. Then the registers A and C are shifted four times. At the end of the first shift, register A contains $(c_3 b_0,\ c_3 b_1,\ c_3 b_2,\ c_3 b_3)$, i.e. the vector representation of $c_3\beta$. At the end of the second shift, register A contains the vector representation of $(c_3\beta)\alpha + c_2\beta$. At the end of the third shift, the content of register A forms the vector representation of $((c_3\beta)\alpha + c_2\beta)\alpha + c_1\beta$. At the end of the fourth shift, register A contains the product $\beta\gamma$ in vector form.

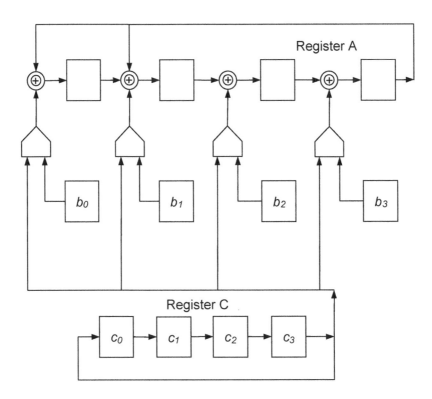

Fig. 6.5. Circuit for the multiplication of two generic elements in $GF(16)$

Also the multiplication of two elements of the Galois field $GF(2^m)$ can be represented by a dedicated table and is implemented as a combinatorial logic with 2^m inputs and m outputs.

Suppose, for example, we want to construct the full multiplier in $GF(2^4)$ with a combinatorial logic.

The multiplication of the two elements follows the scheme represented in Fig. 6.6. First the coefficients of the two elements are multiplied, thus getting 7 coefficients. The coefficients of degree higher than the third, indicated as $AB4$, $AB5$ and $AB6$ are summed to the coefficients of degree less than 4, according to the field rule dictated by Eq. (6.8). To create the combinatorial logic network, we build the matrix that performs the operation. The matrix contains m rows, which indicate the coefficients inside the vector: in other words, the first row indicates the coefficient of degree 0, the second one indicates the coefficient of degree 1 and so on. The columns are m^2: they indicate $a_i b_j$ and are numbered so that the first m columns are $a_0 b_j$, then the second m columns are $a_1 b_j$ and so on up to $a_{m-1} b_{m-1}$.

	α^6	α^5	α^4	α^3	α^2	α^1	$?^0$
α^3	$b_3 a_3$	$b_3 a_2$	$b_3 a_1$	$b_3 a_0$	0	0	0
α^2		$b_2 a_3$	$b_2 a_2$	$b_2 a_1$	$b_2 a_0$	0	0
α^1			$b_1 a_3$	$b_1 a_2$	$b_1 a_1$	$b_1 a_0$	0
α^0				$b_0 a_3$	$b_0 a_2$	$b_0 a_1$	$b_0 a_0$
	$AB6$	$AB5$	$AB4$	0	0	$AB4$	$AB4$
				0	$AB5$	$AB5$	0
				$AB6$	$AB6$	0	0

Fig. 6.6. Multiplication table of two generic elements of $GF(16)$

The matrix contains a 1 where the corresponding coefficient has to be added, otherwise there is a 0. In the example of $GF(2^4)$, the matrix is:

$$\begin{pmatrix} 1 & 0 & 0 & 0 & 0 & 0 & 0 & 1 & 0 & 0 & 1 & 0 & 0 & 1 & 0 & 0 \\ 0 & 1 & 0 & 0 & 1 & 0 & 0 & 1 & 0 & 0 & 1 & 1 & 0 & 1 & 1 & 0 \\ 0 & 0 & 1 & 0 & 0 & 1 & 0 & 0 & 1 & 0 & 0 & 1 & 0 & 0 & 1 & 1 \\ 0 & 0 & 0 & 1 & 0 & 0 & 1 & 0 & 0 & 1 & 0 & 0 & 1 & 0 & 0 & 1 \end{pmatrix} \qquad (6.15)$$

The matrix described by Eq. (6.15) must then be multiplied by the vector c composed of m^2 coefficients as indicated by Eq. (6.16).

$$c = \left(a_0 b_0, a_0 b_1, \ldots, a_0 b_{m-1}, a_1 b_0, \ldots a_1 b_{m-1}, \ldots a_{m-1} b_{m-1}\right) \qquad (6.16)$$

The advantage of this type of implementation is the speed. Anyway, as m increases it becomes prohibitively complex and costly.

In Appendix B (Sect. B.5.2) an implementation of a Galois field multiplication is described.

6.5 Divider

The division is the most complicated operation in a Galois field. Said β and γ two elements of the field, the operation β/γ is equivalent to the multiplication $\beta \gamma^{-1}$.

For a Galois field $GF(q^m)$ the following relation applies:

$$\alpha^{-i} = \alpha^{q^m - i - 1} \qquad (6.17)$$

Thus, for each element it is simple to extract its inverse.

To implement the operation, we can use two different approaches. First of all, the operation may be implemented through a ROM which indicates the inverse of each field element. This "tabular" approach is quite easy if it is done via software, but it becomes prohibitive via hardware, especially in the case of very large field dimensions. In fact, the size of the required ROM is $2^m - 1$ rows and m columns in case of fields $GF(2^m)$.

In the hardware case, it is possible to look for a dedicated combinatorial logic to perform the inversion, in the same way as the multiplier by constant was built. However, also in this case, as the field dimensions increase, the combinatorial logic becomes prohibitive.

When the field dimension is large, we prefer the following approach, by means of successive multiplications. The division is performed with $2m-1$ multiplications which alternate squares and multiplications by constant.

Suppose we want to perform the division of 2 elements a and x in $GF(2^4)$. The division follows Table 6.2 where the left column indicates the multiplication by the divisor x while the right column shows the successive powers to the square.

Table 6.2. Operations to be performed in order to have the inversion in GF(16)

a	a^2
a^2x	a^4x^2
a^4x^3	a^8x^6
a^8x^7	$a^{16}x^{14}$

At the end we get:

$$a^{16}x^{14} = a^{15}a\frac{x^{15}}{x} = \frac{a}{x} \tag{6.18}$$

exploiting the property (Sect. 2.3.1.1):

$$\alpha^{2^m-1} = 1 \tag{6.19}$$

Accordingly, whenever operations are implemented in an extended Galois field, i.e. $GF(2^m)$, which is the interesting case for applicative purposes, we need to keep in mind that the weight of a division is of $2m-1$ multiplications.

In Appendix B (Sect. B.5.3) a routine performing the division is described.

6.6 Linear shift register

Data rates in the hundreds or even thousands of megabits per second are common in many applications. Unfortunately, these date rates limit the technology of the device that may be used for the implementation of error control systems and, having fixed a given technology, they limit the complexity of the circuits. For this reason, it is important to implement the structures as repetitive as possible, by using simple exclusive-OR gates, switches or shift registers. Shift registers (SR) are some of the most simple digital circuits, since they are simply constituted by a collection of flip-flops connected in series. For this reason, given a technology, they can operate at a speed near to the maximum speed possible for a single gate.

In the binary case, shift registers are constituted by binary adders, that is exclusive-OR gates, a memory cell, which is a flip-flop, and multipliers by constant, which are represented either by the existence of a connection (multiplication by 1) or by the absence of a connection (multiplication by 0).

In the non binary case, the operators must be operators of the field $GF(2^m)$. Thus, a shift register is constituted by non binary adders of the type described in Sect. 6.2, by multipliers by constant of the type described in Sect. 6.3 and by memory cells containing more bits. The non binary memory cells are implemented as intuitively expected: a flip flop is devoted to the storage of each coefficient of the vector representation of the field element.

The product of two polynomials $m(x)$ and $g(x)$ may be calculated as a weighted sum of cyclic shifts of $g(x)$, as illustrated in Fig. 6.7.

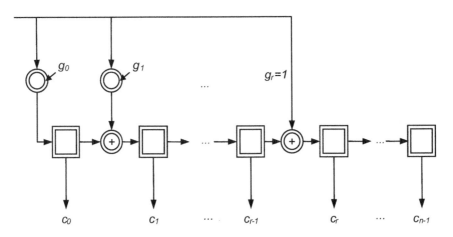

Fig. 6.7. Generic shift register for the multiplication of two polynomials

The coefficients of $m(x)$ are loaded into the circuit in descending order of index. At each clock pulse a new coefficient is set as input and the contents of the memory cells are shifted of one cell to the right. When the final coefficient (m_0) has been loaded into the SR, the memory cells of the circuit contain the result of the polynomial multiplication.

We want, for instance, to calculate the product of $m(x)=1+x^2$ and $g(x)=1+x^2+x^3+x^4$. The circuit that implements the operation is represented in Fig. 6.8.

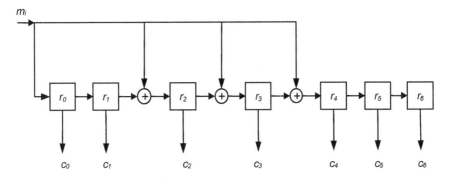

Fig. 6.8. Shift register for the binary multiplication of two polynomials

The contents of the SR cells during the multiplication are represented in Table 6.3.

As we will see later, shift registers are the basic elements encoding and decoding circuits of cyclic codes are constructed with.

Table 6.3. SR cells during the multiplication

SR cells	r_0	r_1	r_2	r_3	r_4	r_5	r_6
Initial state	0	0	0	0	0	0	0
Input $m_2=1$	1	0	1	1	1	0	0
Input $m_1=0$	0	1	0	1	1	1	0
Input $m_0=1$	1	0	0	1	0	1	1
Final state	1	0	0	1	0	1	1

Bibliography

T. C. Chen, S. W. Wei, H. J. Tsai, "Arithmetic Circuit for Finite Field GF(2^M)" in *US Patent 6,687,725 B1*, February 2004.

I. Dror, C. D. Gressel, M. Mostovoy, A. Molchanov, "Extending the Range of Computational Fields of Integers" in *US Patent US2002/0039418 A1*, April 2002.

S. V. Fedorenko, P. V. Trifonov, "Finding Roots of Polynomials Over Finite Fields" in *IEEE Transactions on communications*, Vol. 50, November 2002.

G. L. Fcng, "A VLSI Architecture for Fast Inversion in *GF(2^m)*" in *IEEE Transactions on Computers*, Vol. 38, October 1989.

C. D. Gressel, A. Schevachman, E. Aizman, M. Slobodkin, S. Cooper, "Random Number Slip and Swap Generators" in *US Patent US2004/0205095 A1*, October 2004.

S. Lin, D. J. Costello, "Error Control Coding: Fundamentals and Applications", *F. F. Kuo Editor*, 1983.

M. Metha, V. Parmmar, E. Swartzlander Jr., "High-Speed Multiplier Design Using Multi-Input Counter and Compressor Circuits" in *IEEE Conference on Computer Arithmetic*, 1991.

C. Paar, "Optimized Arithmetic for Reed-Solomon Encoders" in *ISIT*, June 1997.

E. M. Popovici, P. Fitzpatrick, "Division Algorithm over GF(2^m)" in *Electronics Letters*, Vol. 34, September 1998.

M. Potkonjak, M. B. Srivastava, A. P. Chandrakasan, "Multiple Constant Multiplications: Efficient and Versatile Framework and Algorithms for Exploring Common Subexpression Elimination" in *IEEE Transactions on Computers-Aided Design of Integrated Circuits and Systems*, Vol. 15, February 1996.

R. Schober, W. H. Gerstacker, "The Zeros of Random Polynomials: Further Results and Applications" in *IEEE Transactions on Communications*, Vol. 50, June 2002.

S. B. Wicker, "Error Control for Digital Communication and Storage", *Prentice Hall*, 1995.

7 Hamming code for Flash memories

7.1 Introduction

In Chap. 1 we explained how an Error Correction Code can affect the reliability of a Flash memory and Chap. 5 helped us to understand the error sources from which we would like to be protected. Now, after an excursion over the most important error correction codes and over the structure of Flash memories, it is time to go into the implementation of the ECC over the memory.

We will therefore introduce three different situations with the respective theoretical treatment in detail. In the first of these cases a Hamming code is applied to a 2-level NOR memory. In the second case a Hamming code is applied to a 4-level NOR memory and, finally, a Hamming code, for blocks of big dimensions, is applied to a NOR or a NAND memory.

As it is possible to see, the correction of the single error is considered in all these cases. This was the first step in the history of the ECC on Flash memories. Before ECC, physical redundancy was introduced into design to allow the correction of defective parts. This method needs additional circuitry, such as spare columns or spare blocks; anyway, for many years it has been widely accepted as the only way to guarantee reliability in semiconductor industry.

With physical redundancy, the defect detection is made during Electrical Wafer Sort (EWS). If a repairable failure occurs, the redundancy elements are permanently set to replace the defective circuitry.

On the contrary, ECC is able to guarantee a similar repair method, but the detection and correction operations can be performed by the device itself. Moreover, the main difference is that ECC is not constrained to the manufacturing environment and is able to detect and correct defects during the whole device life. Anyway, if a hard fail occurs on a given bit and ECC is able to correct it, it may lose the opportunity to correct another error on another bit. This is the reason why, up to now, physical redundancy has been used in memory devices, in conjunction with ECC.

7.2 NOR single bit

This first application takes into examination the possibility to introduce a Hamming-type ECC for the correction of one error into the structure of an asynchronous NOR-type bi-level Flash memory. The correction of one error in a NOR memory, already very reliable from a structural point of view, takes the residual *BER* of the memory to extremely contained values, as already described in Chap. 1. Such a level of reliability is sufficient in the manufacturing of electro-medical and banking sectors.

Fundamental properties of an asynchronous Flash memory are the area and the access time. If we want to improve the reliability of the memory, by integrating an ECC for the correction of one bit, we must pay special attention to the impact this functionality introduces on these two parameters.

The typical architecture of a modern Flash memory, represented in Fig. 7.1, requires a parallel structure to read data from the matrix. In the represented example, the data are read by the matrix with a parallelism (k) of 128 bits, allowing the implementation of special methods of fast reading, such as the *Asynchronous Page Mode* and the *Synchronous Burst Mode*. These methods provide a tremendous increase in terms of system performances without requiring severe reduction of the access time to the memory. Such techniques are based upon a few considerations about the system, which are briefly detailed here.

Studies and analyses on the correlation of the access to the memory by a large number of different processors that execute a large variety of programs have pinpointed that accesses to contiguous locations are very likely. Intuitively, this is due to the fact that programs are translated into sequences of instructions, stored in neighboring memory locations. This also holds true for both elementary and composed data structures. Furthermore, it must be considered that a large part of the execution time is spent on a limited number of code cycles or functions (about 90% of the time spent on 10% of the code). The foregoing observations, together with the fact that the cycle time of modern processors has been reduced to a few nanoseconds, whereas the access time to an external memory is around some tens of nanoseconds, give origin to system architectures composed of several levels of memory. Typical structures combine a Flash memory of large size, slow and not expensive, with a faster memory. The entire code is copied from the Flash to the internal fast memory, before being executed. In other configurations, one or more levels of very fast cache memories are also present. The inclusion of a cache memory strongly affects the way the processor accesses the memory and the way in which the external memory is used.

Typically, a cache memory is organized as a table that contains a given number of entries. Each entry contains a DATA field and a TAG field. The DATA field is made up of N contiguous words that are aligned with respect to the memory address. The TAG field contains the information about the address to which the data are associated, along with several flags related to the status of validity of the data stored.

Without digging into implementation details, it is important to describe the paged organization of the cache memory. When the data that are required by the processor are not present in the cache, the Memory Management Unit (MMU) generates a signal of "cache miss", providing the selection of one of the pages of the cache that must be substituted for a page coming from the external memory. This means that each access to the external memory is actually carried out by means of a sequence of accesses. Different types of processors adopt different techniques to update the content of the cache. The simplest procedure requires updating the page starting from the first location (aligned mode), and only at the end of the transfer the processor is allowed to continue. The most sophisticated versions demand the immediate load of the word that caused the "miss" signal, freeing the processor from the hold state, and, afterwards, completing the page load (not aligned and wrapped mode). These reasons compel the designer to seek novel Flash architectures that provide higher throughput when sequential transfers of data or pages are required. The Asynchronous Page Mode and the Synchronous Burst Mode are two possible solutions to such a requirement. The Asynchronous Page Mode has a minor impact on the interface toward the system; the Synchronous Burst Mode, which is more complex and requires an external clock, provides the best performances.

In the representation of Fig. 7.1, the basic blocks involved in the asynchronous read phase of a typical NOR Flash memory are shown. The input buffer, address predecoder, ATD block, row and column decoder, cell array, 16-sense amplifier block, output latches, and output buffers form the read path. The ATD block detects the address transitions and triggers the sequence of events. First of all, the predecoder and decoders select the memory cells, then the sense amplifiers are switched on and equalized to read, finally the output latches sample the data, concluding the read operation. The sense amplifiers are then switched off and the array is set in the stand-by condition, while the output buffers are activated to transfer the data to the output pads.

The read access time is determined by the sum of the settling time of the input buffer, the time to read the array, which is controlled by the ATD, and the settling time of the output buffers in the maximum load condition.

In both Page Mode and Burst Mode a parallel data structure is used on the read path. The memory has a paged organization in which each page contains a set of words (each word is made up by 16 bits). Each time the user requests a word, two possibilities may exist.

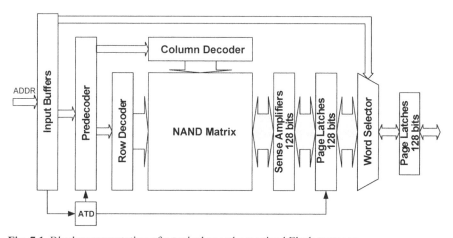

Fig. 7.1. Block representation of a typical paged-organized Flash memory

The first possibility is that the word belongs to a different page from the previous read; the entire page is read from the array and stored in a "page latch". In the second case, the requested word belongs to the previously addressed page; the page latch already contains the requested word that does not need to be read from the array but just selected in the page latch and driven to the output. The address bits (ADDR) have been divided into two groups: the most significant bits select the page to read; the least significant bits are used to select the word inside the page. In Fig. 7.1 a "page latch" stores the result of the read operation and a "word selector" allows us to select the word to present as output on the device pads, according to the word address, i.e. the less significant bits of the entire address.

Having an access point with elevated parallelism in the device, it is easier to introduce the correction functionality with reduced impact, in terms of area, on the storage of the parity bits (Fig. 7.2). The minimum number of parity bits to be used is given by the Hamming inequality (Chap. 1, Eq. (1.28)), that for $q=2$ and $t=1$ is reduced to:

$$2^{n-k} \geq 1 + n \qquad (7.1)$$

For a data field equal to 128 bits (k), the minimum number of bits to be used is equal to 8. Altogether the codeword has a length (n) of 136 bits. The increase of the matrix area is kept down to around 6%. Apart from the matrix area for the parity storage, the ECC requires additional circuits for the reading and the writing of the parity information; it also requires a block which implements the computation of the parity and the correction of the errors.

In order to easily understand the block which implements the ECC (Fig. 7.3), we can theoretically imagine to duplicate the parity calculation block and consider the writing and reading phases in a distinct way, as if we were dealing with a transmission channel (Fig. 7.4).

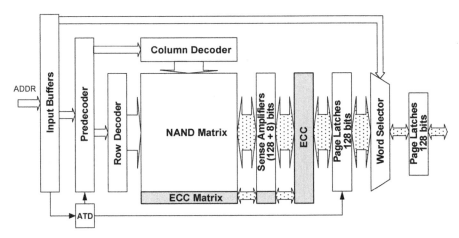

Fig. 7.2. Representation of the architectures of the Asynchronous Page Mode Flash memory with the ECC functionality of Hamming $C[136,128]$ integrated

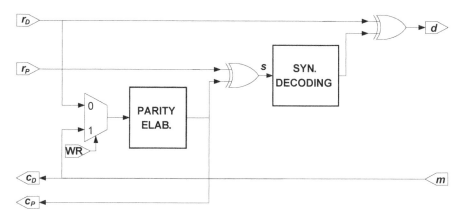

Fig. 7.3. Circuital representation of the function for the calculation of the parity, the decoding and the correction

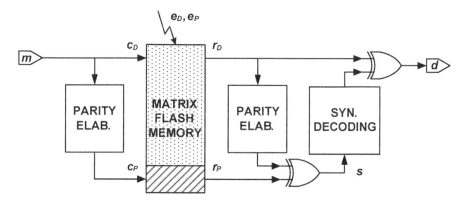

Fig. 7.4. Functional representation of ECC implemented in NOR SLC Flash memory

The vector to be stored in matrix (c) is obtained as the product of the input data vector (m) multiplied by the generator matrix G as described by Eq. (7.2). When working on a binary field, additions and multiplications are translated into operators XOR and AND respectively.

$$c = mG = m(I, P) = (m, mP) = (c_D, c_P) \qquad (7.2)$$

In read, the output vector from the matrix (r) will result to be the same message stored during program, unless some errors occurred. Errors can be represented by an error vector $e=(e_D,e_P)$ and are summed to the written message in accordance with Eq. (7.3).

$$r = c + e \tag{7.3}$$

The syndrome is calculated as:

$$s = rH^T = (c+e)H^T = cH^T + eH^T = mGH^T + eH^T \tag{7.4}$$

Since, by construction, $GH^T=0$, we get that:

$$s = eH^T = (e_D, e_P) \cdot (P^T, I)^T = e_D P + e_P \tag{7.5}$$

that is, the syndrome is function only of the error. Since the encoding is systematic $H=(P^T,I)$, the syndrome (functionally represented in Fig. 7.4) is obtained as in Eq. (7.6).

$$s = (r_D, r_P) \cdot (P^T, I)^T = r_D P + r_P \tag{7.6}$$

Equation (7.6) gives a specific meaning to the rows of the matrix H^T. Suppose we have a single error: if this occurs in the field e_D, in position j, the syndrome assumes the value of the j-th row of P; if the error occurs in the field e_P in position i, the syndrome assumes the value of the i-th row of I. Therefore the matrix $H^T=(P^T,I)^T$ contains the association rule between each syndrome and the respective error position.

The block "Syn Decoding" decodes the syndrome s according to the content of P, generating a vector e_D' that, added to r_D, corrects the error detected. Obviously, it is not useful to detect and to correct errors on the parity bits that are not used externally.

The rows of the matrix P, as already described for every Hamming code, are selected so that:

- all the rows are different from each other;
- all the rows are different from the vector 0;
- all the rows are different from the vectors of I.

The calculation of the parity c_P for the program operation and of the parity $r_D \cdot P$ for the syndrome computation uses the same operator (Parity Elab.) and this is the reason why, in circuital representation (Fig. 7.3), such block is shared by the read path and the write path, depending on the operation being performed.

The parity computation is reduced to a simple product of an input data vector m_j of 128 bits, with a parity matrix P of size 128 × 8. Each of the eight parity bits is calculated as:

$$p_i = \sum_{j=0}^{127} m_j P_{j,i} \qquad (i = 0,\ldots,7) \tag{7.7}$$

From a hardware point of view this means that, for each i-th exit, a tree of XOR sums all the m terms in the locations of the 1s in the P_i column.

In general, the overall complexity of the parity check circuits required by a given codes can be roughly estimated by examining the structure of the parity check matrix P. Basically, the global number of 1s in P determines the complexity of the hardware required: a lower number of 1s requires a less complex circuit. In particular, in systematic codes the total number t_i of 1s in the i-th column of P is related to the number of logic levels necessary to generate the corresponding check bit or syndrome.

Supposing we use a v-inputs mod 2 adder, the number of logic levels required to generate the parity bit p_i and the syndrome s_i is given by:

$$l_{p_i} = \left[\log_v t_i\right] \tag{7.8}$$

$$l_{s_i} = \left[\log_v t_i\right] \tag{7.9}$$

where $[x]$ indicates the smallest integer grater then, or equal to, x.

Finally, to obtain the fastest generation of check and syndrome bits, all t_i should be minimum and equal, or as close as possible, to the average number given by the total number of 1s in P divided by the number of columns.

In Chap. 2 we specified that the Hamming code is a perfect code when used in its canonical form $C[2^b-1, 2^b-b-1]$. Since in our case k is forced to be 128, the minimum useful b is 8. The code used is a shortened Hamming code $C[136, 128]$. The matrix P used is a matrix of size 128×8 bits and is a subset of the parity matrix H^T sized 256×8.

The logical synthesis tools are not very efficient if the choice of the matrix P is not carefully done. Depicted in this way, it is easy to understand that the critical point in the implementation of a Hamming code in a Flash memory is the choice of the parity matrix P.

7.2.1 Generation of the parity matrix

Since such block of parity computation is on the read path, it is important to reduce its crossing time to the minimum. As already said in the previous section, this can be achieved by opportunely choosing the vectors that form the matrix P by minimizing the number of 1s in the matrix.

The problem we are trying to face is to extract a Hamming matrix of dimension $(2^k+b) \times b$ from a complete Hamming matrix of dimension $2^b \times b$. Since we are considering a systematic code, we do not take care of the identity matrix so that the problem becomes the extraction of a parity matrix P of dimension $2^k \times b$ from a matrix $(2^b-b) \times b$. Observe that if b is chosen as the minimum one satisfying Eq. (7.1), then $k=b-1$.

The first step is to verify that

$$\left(\binom{b}{2} + \binom{b}{3} + \dots + \binom{b}{R} \right) > 2^k \qquad (7.10)$$

where R is the highest number of 1s for each row.

The maximum number of 1s for each matrix row is b, while the sum in Eq. (7.10) starts from 2 because this is the minimum number of 1s for each matrix row, since we are not taking care of the identity matrix.

- If the inequality is never verified it is not possible to find the parity matrix P.
- Otherwise we begin to fill up the rows of the matrix with all the vectors with a maximum number of 1s equal to $R–1$.

Now the final choice consists in the extraction of the remaining vectors from a matrix composed by a number Q of 1 for each row where:

$$Q > R - 1 \qquad (7.11)$$

So the real choice is of h elements where

$$h = 2^k - \left[\binom{b}{2} + \binom{b}{3} + \dots + \binom{b}{R-1} \right] \qquad (7.12)$$

After having reduced to the minimum the number of 1s per row, we execute our choice trying to optimize the number of 1s per column. Observe that, contrarily to the row case, it is not useful to reduce to the minimum the number of 1s for each column. The best choice to optimize the columns, according to the previous section, is to try to maximize the number of common terms between them. In this way the total number of adders required for the calculation of the parity can be reduced. Here we try to optimize the number of 1s per column in such a way that, let n be the number of 1s in a column, the following inequality holds:

$$m \le n \le M \qquad (7.13)$$

where m representing the minimum number of 1s is

$$m = \frac{h \cdot R}{b} \qquad (7.14)$$

and M representing the maximum number of 1s is

$$M = \frac{h \cdot Q + b - 1}{b} \qquad (7.15)$$

The drawback of this searching method is that the optimization of the number of 1s made for the columns acts in an independent way from one column to another. In Sect. A.1 we described an example of routine to find a parity matrix for a Hamming code.

This method tries to find a solution of minimal length, i.e. with the use of the minimum number of parity bits. It is possible that the algorithm is not able to find a solution, due to all the imposed constraints. In this case we could relax the constraints regarding the maximum and the minimum number of 1s for each columns. Sometimes this is not the most preferable way and, in applications where the logical critical path in time is considered excessive, it is possible to use more parity bits than the strictly essential ones.

The addition of one more bit to the codewords takes the minimum distance of the Hamming code to 4 (extended Hamming), thus allowing us to identify two errors without performing erroneous corrections, like in the case of the canonical Hamming code. A parity bit calculated as described in Sect. 1.5.1 is the worst possible solution in terms of propagation time and device area, since it depends on all message bits. A more effective method, from this point of view, consists in using the additional bit simply to find an optimal matrix P, adding to the rules of vectors selection for P the condition:

> Each combination of two vectors in P must belong neither to P nor to I; that is each double error must not be confused with a single error.

A constraint commonly required in a code applied to Flash memories is that the parity of a data field of all "1", is all "1". This allows us to have congruent data and parity information immediately after an erase operation. Let w be the parity vector for an input message composed by all "1", such constraint can simply be achieved by adding the vector $I-w$ to the parity, which corresponds to reverse the parity bits into locations where $I-w$ has some "1".

Since it appears twice in the syndrome calculation, the term $I-w$ is simplified.

7.2.2 Generation of the parity matrix P for 2-error Hamming-like codes

In applications requiring the correction of 2 errors, with not very high data length, it is possible to use a Hamming-like approach, instead of the classical 2-error correction codes. The encoding and decoding operations will be exactly the same as for a Hamming code and, again, the critical point is the search of the parity matrix.

Also for this case we will take only systematic codes into consideration. Observe that in this case the minimum distance of the code has to be 5 or 6 and the parity matrix P must contain columns linearly independent for group of 4.

Obviously the first step is to find the minimum number of parity bits required by the Hamming inequality. The algorithmic search is analogous to the previous case, that is we search the parity matrix optimizing the number of 1s.

Let x_i, x_j, x_h and x_k be possible rows of the parity matrix the following conditions must be verified:

$$x_i \neq x_j \quad \forall i, j \tag{7.16}$$

$$x_i + x_k \neq x_j \quad \forall i, j, k \tag{7.17}$$

$$x_i + x_k \neq x_j + x_h \quad \forall i, j, h, k \tag{7.18}$$

In other words this is equivalent to saying:

- every single error cannot be confused with another single error;
- every pair of errors cannot be confused with a single error;
- every pair of errors cannot be confused with another pair of errors.
  Every row is chosen from a matrix systematically sorted as depicted in Fig. 7.5.

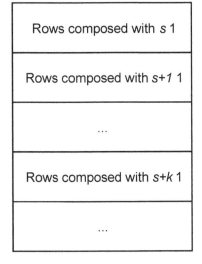

Fig. 7.5. Matrix systematically sorted from which parity vectors are chosen

In this way the chosen matrix results to be sorted as well. This characteristic will be useful during decoding operations.

The encoding operations is performed in the usual way, that is multiplying the message vector by the parity matrix P.

The decoding operation is a little more tricky. The first step finds a syndrome following the flow represented in Fig. 7.4, i.e. the syndrome is the sum between the parity computed on read data and the read parity.

Now the following conditions can occur.

1. The syndrome is a null vector. In this case there were no errors and the read message is a codeword.
2. The syndrome is a vector with one 1. In this case one error occurred and it was on parity bits.
3. The syndrome is a vector with two 1. In this case two errors occurred on parity bits.
4. In all other cases, the syndrome vector has to be searched inside the parity matrix.

First of all we suppose that only one error occurred, so that the syndrome vector s is searched as a row of the parity matrix P.

- If the syndrome corresponds to the i-th row of the matrix, the error occurred at the i-th location.
- If the syndrome is not found as a row of the parity matrix, it can be that two errors occurred and the syndrome correspond to the sum of two rows of the matrix. In this case, being r_i a row of the matrix, the sum indicated in Eq. (7.19) has to be computed

$$syn = s + r_i \quad \forall i \mid r_i \in P \tag{7.19}$$

and the vector syn is searched in the matrix. If the vector syn corresponds to the row r_j than two errors occurred in the positions i and j.

- If the syndrome s is not found neither as a single row of the parity nor as the sum of two rows, then an incorrigible error pattern occurred.

The crucial point in decoding operations is the search of the syndrome in the matrix as a single row or as a sum of two rows.

If the matrix dimensions are too big, this could become a blocking point, since the searching complexity is not linear, but exponential. The parity matrix found with the previous method is sorted by the number of 1s per row. This property can be exploited in the search, because it is possible to use a dichotomic method.

Anyway, it must be said that, even if the sorted generation can give an improvement in decoding algorithm, the computational time, needed by the search, is very high. It can be useful to search the matrix with heuristic method instead of exhaustive ones. This kind of method needs a generator g that must be odd in order to generate the Galois field where the code exists. In this way we build a matrix as in Eq. (7.20).

$$M = \begin{pmatrix} g \\ 2g \\ 3g \\ \dots \\ ng \end{pmatrix} \tag{7.20}$$

The matrix P is chosen as the first rows of M compatible with the constraints of Eqs. (7.16), (7.17) and (7.18).

A possible implementation of this searching routine is described in Sect. A.2.

7.3 NOR Flash multilevel memory

In multilevel memories (MLC) we exploit the ability of the system to check the charge state of the floating gate with great accuracy in order to be able to discriminate, in program and read, q different charge states (Fig. 7.6). In this application we consider a 4-level MLC memory, therefore able to store two information bits per cell.

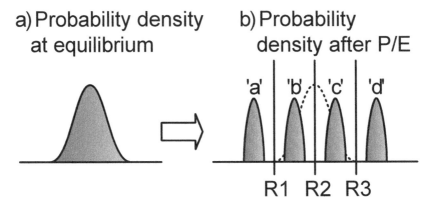

Fig. 7.6. Representation of the four distributions in a 4-level Flash memory

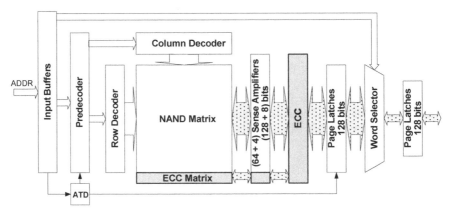

Fig. 7.7. Representation of the architecture of the MLC asynchronous Flash memory with the ECC Hamming functionality $C[68,64]$ integrated in $GF(4)$

The architecture of the memory is similar to the one described for the NOR SLC memory where the parallelism of 128 bits of data is achieved starting from 64 matrix cells through decision circuits (Sense Amplifier) that extract two information bits from each cell (Fig. 7.7).

The minimum number of parity symbols (cells) to be used is given by the Hamming inequality (Eq. (1.28)), that for $q=4$ and $t=1$ is reduced to:

$$4^{n-k} \geq 1 + 3n \qquad (7.21)$$

For k equal to 64 symbols, the minimum number of parity symbols is equal to 4. The used code is therefore a Hamming code $C[68,64]$ in $GF(4)$. All the considerations made for the NOR SLC case can also be applied to this code using operators in $GF(4)$. Since the necessary operations to execute Hamming $C[68,64]$ procedures in $GF(4)$ require operators in $GF(2)$ (XOR and AND), it is useful to translate the code in $GF(2)$. This can simply be obtained by converting the data vectors and the matrices, from $GF(4)$ into $GF(2)$, according to the decimal representation of the elements of $GF(4)$ in $GF(2)$ as described in Table 7.1. For the parity matrix any row pair of $P_{GF(2)}$ corresponds to each row of the matrix $P_{GF(4)}$.

Table 7.1. Decimal conversion from $GF(4)$ towards $GF(2)$

$GF(4)$	0	1	2	3
$GF(2)$	00	01	10	11

7.3.1 Generation of the parity matrix P

In $GF(4)$ it is very easy to get the matrix P. We just have to select 64 different rows, using 4 symbols, different from the null row and different from the identity rows.

Basically, to optimize the critical timing path and the area of the parity calculation block, it is convenient to search $P_{GF(2)}$ as already described for the bilevel NOR case. Being r_i and r_j two rows of the parity matrix, the searching method of the matrix $P_{GF(2)}$ must take care of the following constraints.

- Each row r_i in $P_{GF(2)}$ and the sum of each pair of rows $r_i + r_{i-1}$ with odd i must be different from the vectors with one or two 1.
- Each row r_i in $P_{GF(2)}$ and the sum of each pair of rows $r_i + r_{i-1}$ with odd i must be different from every other row r_j in $P_{GF(2)}$ and from the sum of every other pair of rows $r_j + r_{j-1}$ with odd j, that is the following equations must hold:

$$r_i \neq r_j \qquad (7.22)$$

$$r_i + r_{i-1} \neq r_j \qquad (7.23)$$

$$r_i + r_{i-1} \neq r_j + r_{j-1} \qquad (7.24)$$

- Each row r_i in $P_{GF(2)}$ and the sum of each pair of rows $r_i + r_{i-1}$ with odd i must be different from 0.

In addition to these constraints the usual ones regarding the number of 1s must be verified.

- Minimize the base 2 logarithm of the maximum number of 1s in each column of P, to minimize the number of levels of the adders tree.
- Minimize the number of adders required for the calculation of the parity, by minimizing the number of 1s in the whole matrix P and by maximizing the number of the common terms among the columns.

We leave to the interested reader the exercise of identifying how many cells must be added to take the minimum distance to 4 and to identify the conditions to determine the optimum $P_{GF(2)}$.

Also in the case of double cells correction it is possible to find a parity matrix with exhaustive or heuristic ways. In this case the constraints described in Eqs. (7.22), (7.23) and (7.24) become:

$$r_i \neq r_j \tag{7.25}$$

that is every single error cannot be confused with another single error.

$$r_i + r_{i-1} \neq r_j \tag{7.26}$$

with an odd i. This is equivalent to saying that every erroneous cell cannot be confused with a single bit error.

$$r_i + r_{i-1} \neq r_j + r_{j-1} \tag{7.27}$$

with i and j odd. This is equivalent to saying that every cell error cannot be confused with another cell error.

$$r_i + r_k \neq r_j \tag{7.28}$$

that is every double error cannot be confused with a single error.

$$r_i + r_k \neq r_j + r_h \tag{7.29}$$

that is every double error cannot be confused with another double error.

$$r_i + r_k \neq r_j + r_{j-1} \tag{7.30}$$

with j odd. This is equivalent to saying that is every double error cannot be confused with a single cell error.

$$r_i + r_k \neq r_j + r_{j-1} + r_h + r_{h-1} \tag{7.31}$$

with j and h odd. This is equivalent to saying that is every double error cannot be confused with a double cell error.

$$r_i \neq r_j + r_{j-1} + r_h + r_{h-1} \tag{7.32}$$

with j and h odd. This is equivalent to saying that is every single error cannot be confused with a double cell error.

$$r_i + r_{i-1} \neq r_j + r_{j-1} + r_h + r_{h-1} \tag{7.33}$$

with i, j and h odd. This is equivalent to saying that is every single cell error cannot be confused with a double cell error.

$$r_i + r_{i-1} + r_k + r_{k-1} \neq r_j + r_{j-1} + r_h + r_{h-1} \tag{7.34}$$

with i, j, k and h odd. This is equivalent to saying that is every double cell error cannot be confused with another double cell error. In Sect. A.3 a searching routine for a double cell errors correction is described.

As it possible to see, there are a lot of constraints so that the search could last a very long time. For this reason, when the codeword length of the searched code is too big, it could be better to use the Gray map to find a parity matrix for the double error correction as described above.

7.4 Algorithmic Hamming code for big size blocks

In this section we examine some solutions adopted in systems composed by a microcontroller and by a NOR or a NAND Flash memory, without embedded ECC, and with a reliability considered not sufficient by the designer. In these cases, in order to increase the reliability of the data of the stored code, the controller applies an encoding to the data to be written in the Flash memory and a decoding to the data read and transferred to the system memory. These operations should use the minimum of the system resources, in terms of ROM and RAM memory and in terms of calculation resources (Fig. 7.8).

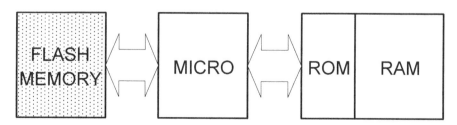

Fig. 7.8. Representation of the reference architecture

Suppose we have an application where the memory is logically divided in 512-Byte blocks and where, in order to reach the required reliability, we want to correct an erroneous bit and to detect two errors. The Hamming inequality (Eq. (1.28)) requires at least 13 bits plus one for the extension of the code. The number of parity bits required is extremely reduced but the parity matrix P consists of 4096 × 14 bits. Unlike the previous applications, where the parity calculation time had to be performed in few nanoseconds and where the data to be decoded were immediately all available, in this application the data are moved from the Flash memory to the data memory byte by byte (and vice versa), so that parity and syndrome computation cannot be immediate. We need a method able to perform the calculation of the parity and of the syndrome without storing the parity matrix in the memory.

As already pointed out

$$s = eH^T = (e_D, e_P)(P^T, I)^T = e_D P + e_P \qquad (7.35)$$

That is the syndrome is a function of only the error. The matrix $H^T = (P^T, I)^T$ contains the information regarding the association rule for each syndrome to the respective error position.

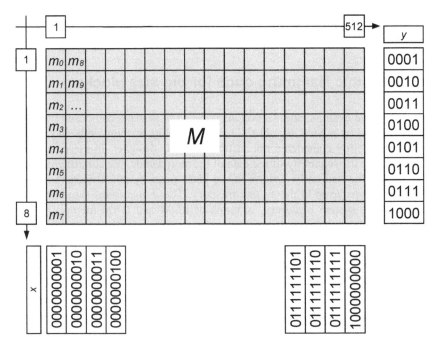

Fig. 7.9. Representation of the matrix M systematically built on the basis of the data flow (m_i) going through the memory. The sorting indexes x (1–512) and y (1–8) univocally identify each bit of the matrix M and are therefore suitable for building the parity matrix H

The flow of 512-Byte data can be represented as a matrix M with size 8×512 (Fig. 7.9). Therefore, the values corresponding to the vectors of the matrix P (Fig. 7.10a) can simply be the coordinates x and y of the position of each bit in the matrix M. Note the lack of the value "0" for the indexes x and y to avoid the null vector and the vectors of the matrix I in P.

Being X and Y the two matrices composing P (Fig. 7.10b), the parity vector p can be calculated as:

$$p = mP = m(Y,X) = (mY,mX) \tag{7.36}$$

$$mX = \sum_{j=1}^{512} j \cdot \sum_{i=1}^{8} M_{ij} \tag{7.37}$$

$$mY = \sum_{i=1}^{8} i \cdot \sum_{j=1}^{512} M_{ij} \tag{7.38}$$

Both the components of the parity vector can be calculated through an algorithmic way with few code lines. The obtained code is a shortened Hamming code $C[4110,4096]$ able to correct one error. Observe that the code, constructed in this way, uses one bit more than the number required by Eq. (1.28). The ability to recognize two errors is reached extending the Hamming code by adding a global parity bit. In the matrix H this bit corresponds to the addition of a column of all "1" (E) and of a null row as shown in Fig. 7.10. The total parity bit can be easily obtained as the parity bit of the sum of all the rows of M.

The syndrome s can be similarly calculated as:

$$s = r_D P + r_P \tag{7.39}$$

With this approach the decoding of s must be performed in an algorithmic way by determining if s belongs to H and what error bit it corresponds to. Being $s=(s_E,s_X,s_Y)$, we follow these steps:

- if the syndrome s is equal to zero, there are no errors;
- if the bit $s_E=1$ and (s_X, s_Y) identify one bit in the matrix M (that is $s_X \in \{1,...,8\}$ and $s_Y \in \{1,...,512\}$), only one error is detected, in position s_X, s_Y;
- if the bit $s_E=1$ and $(s_X, s_Y) \in I$, the data matrix M is recognized as correct and one error is detected on the parity bits;
- in any other case two or more errors are detected.

The implementation of the encoding and the decoding operations requires few code lines and every byte requires two or three operations for the execution. Moreover, with this encoding technique, it is possible to have that the parity of all "1" is all "1", simply by adding $1-w$ to the parity obtained, where w is the parity of a sequence of all "1". Another simple way to satisfy this constraint is to calculate the parity of the logical inversion of *M*, since the parity of the null message is the codeword zero.

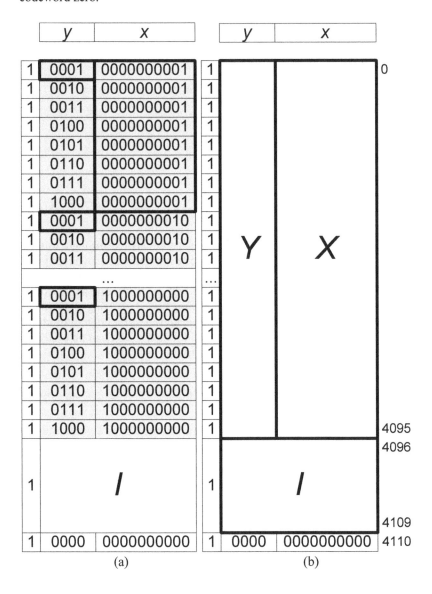

(a)   (b)

Fig. 7.10. Representation of the matrix *H*. The sorting indexes *x* (1–512) and *y* (1–8), univocally identify each row of the matrix *H* (**a**). For the calculation of the parity and of the syndrome, the matrix *H* is seen as the composition $((Y,X)^T,I)$. The matrix *H* is completed by adding the necessary row and column for the code extension

The code can easily fit the microcontroller structures using words' width different from the eight bits case described here.

A similar code for NAND applications is described by Samsung[1] but it is absolutely general. The difference, compared to what illustrated before, is in the matrix P obtained starting from two sequences x and y sorted from 0 to 511 for the x and from 0 to 7 for the y (Fig. 7.11). The matrix P is obtained through the composition of X, X', Y, Y' as represented in Fig. 7.12a. Observe that every vector in x' is the logical inversion of the corresponding x vector and every vector y' is the logical inversion of the corresponding y vector. In this way all the constraints are satisfied so that P is a matrix for a Hamming code: in fact all the rows are unique and are equivalent neither to the vector 0 nor to the vectors of I. By construction, the code is able to detect two errors because, as it can be shown, the sum of two rows of P never belongs to P and it is always different from a vector of I. In fact, the sum of two vectors of P, e.g. (a,a') and (b,b') always originates a vector of type (c,c) which belongs neither to I nor to P.

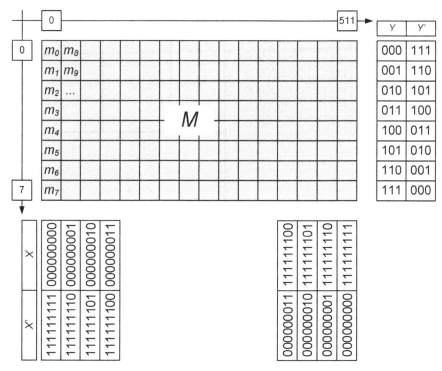

Fig. 7.11. Representation of the matrix M systematically built on the basis of the data flow going through the memory. The sorting indexes x (0–511) and y (0–7) univocally identify each bit of the matrix M and are therefore suitable for building the parity matrix H

---

[1] "Application note for NAND Flash Memory", Samsung Electronics, 1999, www.samsung.com

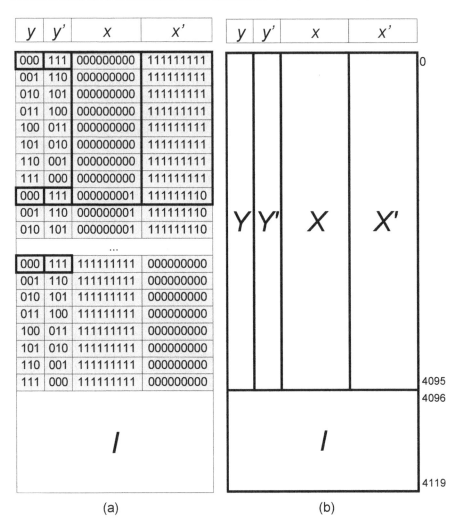

(a)   (b)

Fig. 7.12. Representation of the matrix H. The sorting indexes x (0–511) and y (0–7), univocally identify each row of the matrix H (**a**). For the parity and the syndrome calculation, the matrix H is seen as the composition $((Y,Y',X,X')^{T},I)$ (**b**)

Being X, X', Y and Y' the four matrices composing P (Fig. 7.12b), the parity vector p can be calculated as:

$$p = mP = m(Y,Y',X,X') = (mY,mY',mX,mX')$$ (7.40)

where

$$mX = \sum_{j=0}^{511} j \cdot \sum_{i=0}^{7} M_{ij} \qquad (7.41)$$

$$mX' = \sum_{j=0}^{511} j' \cdot \sum_{i=0}^{7} M_{ij} \qquad (7.42)$$

$$mY = \sum_{i=0}^{7} i \cdot \sum_{j=0}^{511} M_{ij} \qquad (7.43)$$

$$mY' = \sum_{i=0}^{7} i' \cdot \sum_{j=0}^{511} M_{ij} \qquad (7.44)$$

Given the syndrome $s = (s_X, s_{X'}, s_Y, s_{Y'})$, the relationships to be verified during decoding operations can be summarized as:

- if $s=0$, there are no errors;
- if $s \in P$ and therefore $s_X = (s_{X'})$ and $s_Y = (s_{Y'})$, the code detects the bit $M_{sx,sy}$ as the erroneous one;
- if $s \in I$, the data in M are correct and one error is detected on the parity bit;
- in any other case two or more errors are detected.

This method is even more regular than the previous one: it does not require a comparison of the indexes and is therefore more suitable to a hardware implementation.

Usually if the correction of more than one bit is needed, classical methods are preferred to the matrix search. This is particularly true because we do not want to store the parity matrix and we need a method able to treat a big data quantity.

In the next two chapters these methods will be described in details.

Bibliography

C. L. Chen, "Symbol Error-Correcting Codes for Computer Memory Systems" in *IEEE Transactions on Computers,* Vol. 46, February 1992.

C. L. Chen, "Symbol Error Correcting Codes for Memory Applications" in *IEEE Proceedings of FTCS,* 1996.

C. L. Chen, M. Y. Hsiao, "Error-Correcting Codes for Semiconductor Memory Applications: A State-of-the-Art-Review" in *IBM Journal of Research and Development,* March 1984.

A. A. Davydov, L. M. Tombak, "An Alternative to the Hamming Code in the Class of SEC-DED Codes in Semiconductor Memory" in *IEEE Transactions on Information Theory*, Vol. 37, May 1991.

I. Dror, M. Avraham, B. Dulgunov, E. Fumbarov, "Compact High-Speed Single-Bit Error-Correction Circuit" in *US Patent US2005/0160350 A1*, July 2005.

F. Fummi, D. Sciuto, C. Silvano, "Automatic Generation of Error Control Codes for Computer Applications" in *IEEE Transactions on Very Large Scale Integration (VLSI) Systems*, Vol. 6, July 1998.

S. Gregori, A. Cabrini, O. Khouri, G. Torelli, "On-Chip Error Correcting Techniques for New-Generation Flash Memories" in *Proceedings of the IEEE*, Vol. 91, April 2003.

S. Gregori, O. Khuori, R. Micheloni, G. Torelli, "An Error Control Code Scheme for Multi-level Flash Memories", *2001 IEEE International Workshop on Memory Technology, Design and Testing (MTDT 2001), San Jose, California (USA)*, August 2001, pp. 45–49.

M. Harmada, E. Fujiwara, "A Class of Error Control Codes for Byte Organized Memory Systems - SbEC-(Sb+S)ED Codes" in *IEEE Transactions on Computers*, Vol. 46, January 1997.

T. Ken, T. Tomoharu, T. Toru, "A Multipage Cell Architecture for High-Speed Programming Multilevel NAND Flash Memories" in *IEEE Journal of Solid State Circuits*, Vol. 33, August 1998.

M. Lasser, "Method of Managing a Multi-Bit-Cell Flash Memory" in *US Patent US2006/0004952 A1*, January 2006.

P. Makumber, "Design of a Fault-Tolerant Three-Dimensional Dynamic Random-Access Memory with On-Chip Error-Correcting Circuit" in *IEEE Transactions on Computers*, Vol. 42, December 1993.

P. Mazumder, "Design of a Fault-Tolerant Three-Dimensional Dynamic Random-Access Memory with On-Chip Error-Correcting Circuit" in *IEEE Transactions on Computers*, Vol. 42, December 1993.

D. N. Nguyen, S. M. Guertin, A. H. Johnston, "Radiation Effects on Advanced Flash Memories" in *IEEE Transactions on Nuclear Science*, Vol. 46, December 1999.

E. Ou, W. Yang, "Fast Error-Correcting Circuits for Fault-Tolerant Memory", in *IEEE International Workshop on Memory Technology, Design and Testing*, 2004.

L. Penzo, D. Sciuto, C. Silvano, "Construction Tecniques for Systematic SED-DED Codes with Single Byte Error Detection and Partial Correction Capability for Computer Memory Systems" in *IEEE Transactions on Information Theory*, Vol. 41, March 1995.

B. Polianskikh, Z. Zilic, "Design and Implementation of Error Detection and Correction Circuitry for Multilevel Memory Protection" in *Proceedings of the 32nd IEEE International Symposium on Multiple-Valued Logic*, 2002.

D. Rossi, C. Metra, B. Riccò, "Fast and Compact Error Correcting Scheme for Reliable Multilevel Flash Memories" in *IEEE International On-Line Testing Workshop*, 2002.

M. Spica, "Do We Need Anything More than Single Bit Error Correction (ECC)?" in *IEEE International Workshop on Memory Technology, Design and Testing*, 2004.

T. Tanzawa, T. Tanaka, K. Takeuchi, R. Shirota, S. Aritome, H. Watanabe, G. Hemink, K. Shimizu, S. Sato, Y. Takeuchi, K. Ohuchi, "A Compact On-Chip ECC for Low Cost Flash Memories" in *IEEE Journal of Solid-State Circuits*, Vol. 32, May 1997.

T. Toru, T. Ken, S. Riichiro, A. Seiichi, W. Hiroshi, H. Gertjan, S. Kazuhiro, S. Shinji, T. Yuji, O. Kazunori, "A Compact On-Chip ECC for low Cost Flash Memories" in *IEEE Journal of Solid State Circuits*, Vol. 32, May 1997.

8 Cyclic codes for non volatile storage

8.1 General structure

The cyclic codes considered in this chapter are the primitive narrow-sense binary BCH codes and the Reed-Solomon codes introduced in Chap. 2. In describing how the choice, the encoding and the decoding operations are made, the explanation will proceed in parallel on both these classes of codes.

First of all, we will explain how the parameters of a BCH code and a Reed-Solomon one are chosen, considering those cases which are interesting for practical applications, that is to say binary codes for the BCH class and over $GF(2^m)$ for the Reed-Solomon one.

8.1.1 Parameters for the choice of a binary BCH code

As explained in Chap. 2, the description of an error correction code starts by setting the length of the codeword n. For a code $BCH[n,k]$ such length is theoretically determined as $n=2^m-1$. Setting n means setting the length of the codeword composed by the information bits and by the parity ones. In applications this length does not always coincides with the one required by the theory, as the design of a code starts by setting the length k and the correction capability t that we want to get. However, as already seen in Chap. 1, this does not represent a limit, since it is possible to perform the shortening operation.

A shortened cyclic code is generally not cyclic anymore but it is possible to design it in the same way.

Since, in applications, we usually have $k=2^s$ the description will proceed referred to this case.

First of all, we have to understand over what field $GF(2^m)$ we are operating on, by solving the following inequality where m is unknown:

$$2^m - 1 \geq 2^s + mt \qquad (8.1)$$

The solution for m enables us to know how much parity mt is needed in order to correct each combination of t errors at the most.

Suppose, for example, we want to correct 4 errors on a message that carries 512 information bits.

Equation (8.1) is reduced to:

$$2^m - 1 \geq 512 + 4m \tag{8.2}$$

The minimum m that verifies the inequality is 10, consequently the code operates over $GF(2^{10})$.

The whole structure will therefore be described considering a $BCH[2^m-1,2^m-1-mt]$ code, subsequently shortened to get a code $BCH[2^s+mt,2^s]$.

In the above example we construct the code $BCH[1023,1023-40]$, which shortened will become $BCH[512+40,512]$.

It will now be explained how the construction of a BCH code proceeds. First of all remember that, for every Galois field, a primitive polynomial (Sect. 2.3.1.1), which has to be identified in order to perform the necessary operations, is defined.

The primitive polynomials of minimum weight are well known and tabulated[1] up to $GF(2^{32})$. Once the working field of the code is described, the search for the t minimal polynomials enables us to entirely describe the code, as the generator polynomial of the code is the product of the t minimal polynomials.

By construction, a primitive narrow-sense binary BCH code needs the minimal polynomials of α, α^3, α^5,..., α^{2t-1}. Note that the minimal polynomial of α coincides with the primitive polynomial of the field. The minimal polynomials are tabulated[1,2] up to $GF(2^{10})$; for bigger dimensions we need a method to find them. Special mathematical softwares, like Matlab, facilitate the task, as there already exist partially implemented functions able to calculate them. In Sect. B.1 there is a simple implementation of a Matlab function which determines the minimal polynomials and the generator polynomial of a BCH code whose dimension and correction capability have been set. There are also constructive methods to find them, based on cyclotomics cosets, which are not the aim of the present book[1,3].

At this point the parameters of a BCH code have been defined. The general structure is represented in Fig. 8.1.

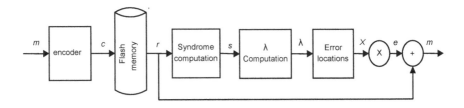

Fig. 8.1. Structure of a binary BCH code

[1] S. B. Wicker, "Error Control for Digital Communication and Storage", *Prentice Hall*, 1995.
[2] S. Lin, D. J. Costello, "Error Control Coding: Fundamentals and Applications", *F. F. Kuo Editor*, 1983.
[3] T. Moon, "Error Correction Coding – Mathematical Methods and Algorithms", *John Wiley and Sons*, 2005.

The encoding consists of only one block, as it is represented by only one operation; contrarily the decoding is composed of 3 steps, as described in Chap. 2: for this reason it turns out to be much more onerous in terms of calculation and latency. In the following, we describe the implementation of these steps, the complexity and the possible simplifications.

8.1.2 Parameters for the choice of a Reed-Solomon code

Similarly to BCH codes, also for Reed-Solomon codes the starting point of the theory does not always fit the starting point of the applications.

In fact, also in this case, the first parameter fixed by the theoretical description concerns the length $n=2^m-1$. For Reed-Solomon codes, however, the length indicates the total number of symbols of m bits: such number is composed of information symbols and $2t$ parity symbols. On the contrary, in applications, the starting point for the design of a code is the information length and the correction capability t.

As for BCH codes, supposing we have a message of 2^s bits to encode, the first step consists in understanding in what field $GF(2^m)$ we are operating on. We get the answer by solving the following inequality where m appears as unknown:

$$m\left(2^m - 1\right) \geq 2^s + 2mt \tag{8.3}$$

Note that the code corrects t symbols of m bits each.

Returning to the example already considered for BCH codes, suppose we want to correct 4 errors on a message that carries 512 information bits, this time with a Reed-Solomon code.

Equation (8.3) is reduced to:

$$m\left(2^m - 1\right) \geq 512 + 8m \tag{8.4}$$

The minimum value of m that solves the equation is $m=7$. The code will then exists over $GF(2^7)$.

The shortening operation can be performed also on Reed-Solomon codes, keeping in mind that a shortened Reed-Solomon code is not cyclic anymore and symbols, instead of bits, are removed.

The whole structure will therefore be constructed considering a Reed-Solomon code $RS[127,119]$, subsequently shortened to get a code $RS[82,74]$. The shortening operation for Reed-Solomon codes does not always allow us to get exactly the k required by the application. In the example, the shortened code carries $74 \times 7 = 518$ information bits, 6 more then those required. For this reason the first symbol will be therefore composed of 6 zeroes and only one information bit.

Note that the value of m found by solving Eq. (8.3) is a minimum value: in fact in some applications it may be useful to use values larger then m. Suppose, for example, we want to correct 4 errors on a message, whose information bits are 4096. Equation (8.3) becomes:

$$m\left(2^m - 1\right) \geq 4096 + 8m \tag{8.5}$$

The minimum value of m that verifies the inequality is 9. For this reason we could construct a code over $GF(2^9)$ of the type $RS[511,503]$ subsequently shortened to get a code $RS[464,456]$. Often, in real applications, the memory manages a flow of bytes before and after the error correction code. Consequently, the preferable Reed-Solomon code is the one able to deal with the same quantities of the memory, that is to say bytes. As seen, this it is not always possible, therefore the code presented in this example would require a structure of the type represented in Fig. 8.2.

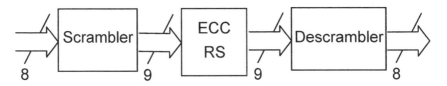

Fig. 8.2. Structures that are linked to ECC

Before and after the ECC machines, i.e. encoder and decoder, there are some structures indicated as Scrambler and Descrambler which turn the 8-bit data bus required for storage operations into a 9-bit bus necessary for the ECC and, vice versa, the 9-bit data bus outgoing the ECC into an 8-bit bus available for the memory.

An alternative solution can be the following. We want to turn the 9-bits operating code into a code that can easily process a byte. Suppose we do not choose the minimum m of Eq. (8.5) but we choose $m=10$. Each symbol formed by 10 bits will be composed of 8 data bits and 2 bits of 0s. In this way the code works and it is constructed over $GF(2^{10})$ but, somehow, it operates directly on bytes. The code will become a code $RS[1023,1015]$ subsequently shortened to $RS[520,512]$ with symbols of 10 bits.

With this procedure we do not need the scrambler and descrambler structures anymore. Obviously, we need an evaluation of the real cost of this type of implementation: in fact, the operators will be a bit more complicated and the ECC structures must identify the last two bits of the symbol as identically zeroes.

Finally, note that the BCH codes presented above do not need scrambler and descrambler structures, as they are serial codes that deal with only one bit at a time.

Similarly to BCH codes, also RS codes are described through a generator polynomial. As seen in Chap. 2 (Sect. 2.3.3), for a Reed-Solomon code it is very simple to calculate it, because the minimal polynomial of α over $GF(2^m)$ is $(x-\alpha)$. It is then sufficient to choose the product of $2t$ terms $(x-\alpha^i)$, with $0 \leq i \leq 2t-1$.

The general structure of a Reed-Solomon code is represented in Fig. 8.3.

The encoding is composed of a single operation, while the decoding is composed of 4 steps and it results to be more onerous, in terms of complexity and latency, in comparison with a BCH code.

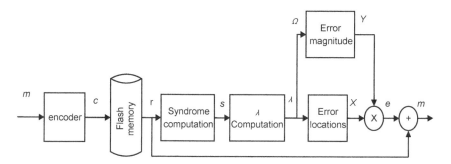

Fig. 8.3. Structure of a Reed-Solomon code

8.2 Encoding

8.2.1 Binary BCH

The encoding of a systematic BCH code is performed by multiplying the message $m(x)$ by x^{n-k} and calculating the parity bits as the remainder of the division of this multiplication by the generator polynomial, in accordance with Eqs. (8.6), (8.7) and (8.8).

$$\frac{m(x) \cdot x^{n-k}}{g(x)} = q(x) + \frac{r(x)}{g(x)} \tag{8.6}$$

$$m(x) \cdot x^{n-k} + r(x) = q(x) \cdot g(x) \tag{8.7}$$

$$c(x) = m(x) \cdot x^{n-k} + r(x) \tag{8.8}$$

Software encoding performs the operations pinpointed in Eqs. (8.6), (8.7) and (8.8) according to the Galois field rules as described in Chap. 6.

However, the best way to perform the encoding process is through hardware implementation, since the operation has to be fast and the structure is very regular.

First of all, note that multiplying $m(x)$ by x^{n-k} is equivalent to simply shifting the message to the right of $n-k$ positions. Then, the parity bits are calculated as the remainder of a polynomial division. The structure that calculates this remainder is based on a linear shift register.

Hardware encoders work on pipelined data, therefore the calculation of the division uses the bits of the message one at a time. We will use an example to describe how this happens.

Suppose to have a code $BCH[15,7]$ whose generator polynomial is $g(x) = x^8+x^7+x^6+x^4+1$ and to encode the message $m(x)=x^6+x^5+x^3+x^2+ 1=(1101101)$.

The multiplication by x^{n-k} is used only to locate the message in the most significant bits of the encoded word, while the polynomial division calculates in k steps the parity bits located in the position of the least significant bits.

The divisor is $g(x)$, while the dividend is $x^{n-k} m(x)$. The division follows the structure of Table 8.1.

Table 8.1. Structure of a polynomial division

		x^{14}	x^{13}	x^{12}	x^{11}	x^{10}	x^9	x^8	x^7	x^6	x^5	x^4	x^3	x^2	x^1	x^0
			0	0	0	0	0	0	0	0						
1	→	1														
	g ·	1	1	1	0	1	0	0	0	1						
			1	1	0	1	0	0	0	1						
1	→		1													
	g ·		0	0	0	0	0	0	0	0	0					
				1	0	1	0	0	0	1	0					
0	→			0												
	g ·			1	1	1	0	1	0	0	0	1				
					1	0	0	1	0	1	0	1				
1	→				1											
	g ·				0	0	0	0	0	0	0	0	0			
						0	0	1	0	1	0	1	0			
1	→					1										
	g ·					1	1	1	0	1	0	0	0	1		
							1	0	0	0	0	1	0	1		
0	→						0									
	g ·						1	1	1	0	1	0	0	0	1	
								1	1	0	1	1	0	1	1	
1	→							1								
	g ·							0	0	0	0	0	0	0	0	0
									1	0	1	1	0	1	1	0

The registers are initialized with zeroes. The bits of the message must be considered from the most significant to the least significant one and they are represented in the column to the left.

The first bit of the message "1" is added to the content of the most significant column. The result "1" is multiplied by the coefficients of the generator polynomial in order to have the values to be added to the content of the remaining columns. When the last bit of the message enters the structure and the last sum is done, the last 8 registers contain the remainder of the division.

The structure that implements this division is represented in Fig. 8.4.

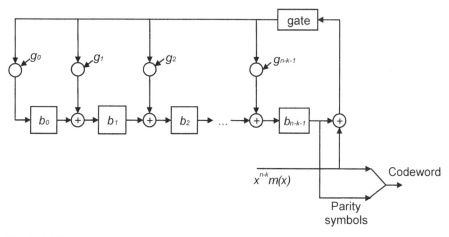

Fig. 8.4. Binary BCH encoder

Note that, since we are considering binary BCH codes, the sum is actually a XOR, while the product is an AND.

The structure is formed by $n–k$ registers and there can be a maximum number of n-k multiplications by the coefficients of the generator polynomial. The content of the registers has to be summed to the incoming bit if the corresponding coefficient of the generator polynomial is 1, otherwise there will not be any wire indicating the sum.

For the first k steps, the message inputs the structure and contemporarily goes out to form the codeword. At the end, the $n–k$ registers contain the parity bits. The parity bits output in $n–k$ clock pulses. Note that the latency of this structure is n.

Shortening means adding many 0s in the most significant bits of the message in order to reach the length required by the theory of $2^m–1–mt$. In the above structure the most significant bits, that is these added 0s, are the first ones to input, so that the content of the registers does not change.

Accordingly, the shortening does not have any effect on this structure, since the latency remains the real length n and the parity bits are calculated in $k=2^s$ clock pulses.

In Appendix B (Sect. B.2) a C-language routine performing the encoding operations for a BCH is described.

8.2.2 Reed-Solomon

The encoding operations for a Reed-Solomon code are described by the same equations of a BCH code (Eqs. (8.6), (8.7) and (8.8)); the only difference is that we are not working in a binary field anymore, but over a field of type $GF(2^m)$.

A software implementation executes the operations sequence as described by the equations; anyway, using the same considerations given for the BCH code, also in this case the best implementation is the hardware one.

The Reed-Solomon code works on symbols and not on bits, therefore the message to be encoded is formed by a sequence of k symbols of s bits. The first step turns out to be a structure that groups the bits flow into symbols before starting the encoding process.

Also in this case the division is pipelined in a sequential way on symbols. With reference to Table 8.2, the division is applied to the example of a Reed-Solomon code *RS[15,11]* whose generator polynomial is $g(x)=x^4+15x^3+3x^2+x+12$. The tables for the sum and multiplication over $GF(2^4)$ are in Appendix C and are calculated as described in Chap. 6.

Suppose we want to encode the message $(1,2,3,4,5,6,7,8,9,10,11)$: the encoding structure is still based on a LSR which performs the polynomial division.

The structure (Fig. 8.5) is analogous to the one depicted for binary BCH codes; the only differences belong to the working field. We have $2t$ registers of s bits, which will have to contain the parity symbols and $2t$ multiplications by the coefficients of the generator polynomial. In contrast with binary BCH codes, where these multiplications were a logical AND between bits, here there are $2t$ multiplications by constant in $GF(2^m)$ that will be performed according to the rules described in Chap. 6.

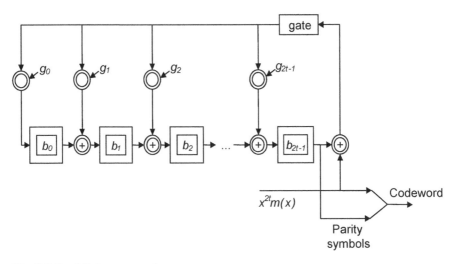

Fig. 8.5. Reed-Solomon encoder

The parity symbols are calculated in k clock pulses and then other $2t$ clock pulses are needed to extract the parity symbols contained in the registers. The latency to encode a message of ks bits is n.

The effect of the shortening operation is analogous to what said for BCH codes, keeping in mind that for Reed-Solomon codes the shortening acts on symbols instead of bits, so there may be a symbol partially filled in with 0s, as explained in Sect. 8.1.2.

Table 8.2. Structure of polynomial division for a Reed-Solomon code

		x^{14}	x^{13}	x^{12}	x^{11}	x^{10}	x^9	x^8	x^7	x^6	x^5	x^4	x^3	x^2	x^1	x^0
		0	0	0	0											
1	$\rightarrow$	1														
	$g \cdot$	1	15	3	1	12										
			15	3	1	12										
2	$\rightarrow$		2													
	$g \cdot$		13	7	4	13	3									
				4	5	1	3									
3	$\rightarrow$			3												
	$g \cdot$			7	11	9	7	2								
					14	8	4	2								
4	$\rightarrow$				4											
	$g \cdot$				10	12	13	10	1							
						4	9	8	1							
5	$\rightarrow$					5										
	$g \cdot$					1	15	3	1	12						
							6	11	0	12						
6	$\rightarrow$						6									
	$g \cdot$						0	0	0	0	0					
								11	0	12	0					
7	$\rightarrow$							7								
	$g \cdot$							12	8	7	12	15				
									8	11	12	15				
8	$\rightarrow$								8							
	$g \cdot$								0	0	0	0	0			
										11	12	15	0			
9	$\rightarrow$									9						
	$g \cdot$									2	13	6	2	11		
											1	9	2	11		
10	$\rightarrow$										10					
	$g \cdot$										11	3	14	11	13	
												10	12	0	13	
11	$\rightarrow$											11				
	$g \cdot$											1	15	3	1	12
													3	3	12	12

8.3 Syndromes calculation

8.3.1 Binary BCH

The first step of the decoding procedure consists in calculating the syndromes.

As already explained in Chap. 2, the calculation of the syndromes allows us to understand if some errors have occurred, as identically null syndromes mean that the read message is correct. In such case we do not proceed with the decoding, which turns out to be very fast.

Since we are considering binary codes, we need to calculate only t syndromes, that is, those linked to the odd powers of the primitive element.

Remembering what said in Chap. 2, the calculation of the syndromes is done through Eqs. (8.9), (8.10) and (8.11).

$$\frac{R(x)}{\psi_i(x)} = Q_i(x) + \frac{S_i(x)}{\psi_i(x)} \tag{8.9}$$

$$S_i(x) = Q_i(x) \cdot \psi_i(x) + R(x) \tag{8.10}$$

$$S_i = R(\alpha^i) \tag{8.11}$$

Therefore there are two methods to effectively calculate the syndromes:

- as remainders of the division between the received polynomial and the minimal polynomial ψ_i, then evaluated in α^i;
- as evaluation of the polynomial received in α^i.

The best method depends on whether the selected implementation is hardware or software.

The implementation via hardware is very useful if the structure is very repetitive and regular. In this case the implementation could result in a circuit that performs all the calculations very fast. The drawback of a hardware implementation is the occupied area. If there is the need to store a lot of information, the hardware implementation is not suitable because we need a lot of latches or flip-flops.

On the other side, a software implementation is useful to perform a lot of calculations, especially if there are a lot of different calculations that are neither repetitive nor regular. The software does not have area constraints: so if it is necessary to store a lot of information the software is the best choice. From the speed point of view, the software has the constraint of the CPU frequency that is the number of instructions in a unit time.

For these reasons, the software calculation has a preference for the second method and the substitution is performed through the so-called Horner rule, represented by Eq. (8.12).

$$S_i = R(\alpha^i) = R_{n-1}(\alpha^i)^{n-1} + R_{n-2}(\alpha^i)^{n-2} + \ldots + R_1\alpha^i + R_0$$

$$= (\ldots((R_{n-1}\alpha^i + R_{n-2})\cdot\alpha^i + R_{n-3})\cdot\alpha^i + \ldots + R_1)\cdot\alpha^i + R_0 \tag{8.12}$$

This calculation requires $n-1$ sums and $n-1$ products. In total, for t syndromes we need $(n-1)t$ sums and $(n-1)t$ products.

Instead, if the calculation is done via hardware the first method can be used. In this case the structure is entirely analogous to the encoder and is based on a linear shift register, since, in both cases, we have to calculate the remainder of a polynomial division.

The syndromes calculation for a BCH code existing over $GF(2^m)$ involves t structures which contemporarily calculate the remainder of the divisions between the received polynomial and the t minimal polynomials.

Each of these structures has m registers and there are m sums at the most. The sums, which are XOR in this case, take place where the corresponding coefficient of minimal polynomial is a 1. We need n clock pulses to have these remainders in the registers.

At this point the calculation is not finished yet, since it is necessary to evaluate these remainders in the α powers.

Evaluating a polynomial in α^i means substituting α^i in the polynomial remainder $r(x)$. Note that the substitution of α does not involve any change in the corresponding remainder; this is equivalent to saying that, for narrow-sense primitive codes, the registers of the first structure already contain the first syndrome and it is not necessary to perform any evaluation.

For the other structures we have:

$$S_i = r(\alpha^i) = r_{m-1}(\alpha^i)^{m-1} + \ldots + r_1\alpha^i + r_0$$

$$= r_{m-1}\alpha^{im-i} + \ldots + r_1\alpha^i + r_0 = c_{m-1}\alpha^{m-1} + \ldots + c_1\alpha + c_o \tag{8.13}$$

The calculation of the coefficients c_i depends on the field we are working on.

Suppose, for example, we are working over $GF(16)$, where the primitive polynomial is $1+x+x^4$. Called α the primitive element, we have $\alpha^4 = \alpha+1$. With reference to Appendix C, it is possible to express all the powers of α as polynomials of third degree at the most.

We want, for example, to calculate the second syndrome. Suppose that the division between the received polynomial $R(x)$ and the minimal polynomial ψ_3 has given the remainder $r(x)=r_0+r_1x+r_2x^2+r_3x^3$ as a result. The calculation of the syndrome is complete once we have evaluated this polynomial in α^3. We have:

$$S_3 = r_0 + r_1\alpha^3 + r_2(\alpha^3)^2 + r_3(\alpha^3)^3 = r_0 + r_1\alpha^3 + r_2\alpha^6 + r_3\alpha^9$$

$$= r_0 + r_1\alpha^3 + r_2(\alpha^3 + \alpha^2) + r_3(\alpha^3 + \alpha)$$

$$= r_0 + r_3\alpha + r_2\alpha^2 + (r_1 + r_2 + r_3)\cdot\alpha^3 \tag{8.14}$$

Accordingly, the coefficients s_i of the syndrome are calculated from the remainder as:

$$s_0 = r_0$$
$$s_1 = r_3$$
$$s_2 = r_2 \tag{8.15}$$
$$s_3 = r_1 + r_2 + r_3$$

It is therefore a logic network to be applied to the remainder in order to have the right coefficients and can be represented by a matrix $m \times m$. In this case the evaluation is performed in only one clock pulse.

Each evaluation is represented by a matrix $m \times m$. The result of the multiplication between the i-th polynomial remainder and the corresponding matrix is the polynomial remainder evaluated in α^i.

In the example described above the evaluation matrix for the second syndrome is:

$$M = \begin{pmatrix} 1 & 0 & 0 & 0 \\ 0 & 0 & 0 & 1 \\ 0 & 0 & 1 & 1 \\ 0 & 1 & 0 & 1 \end{pmatrix} \tag{8.16}$$

$$(s_0, s_1, s_2, s_3) = (r_0, r_1, r_2, r_3) \cdot \begin{pmatrix} 1 & 0 & 0 & 0 \\ 0 & 0 & 0 & 1 \\ 0 & 0 & 1 & 1 \\ 0 & 1 & 0 & 1 \end{pmatrix} \tag{8.17}$$

In Sect. B.4 a Matlab routine that finds these matrices is described.

The structure of the syndromes calculation for a code $BCH[15,7]$ corrector of two errors is represented in Fig. 8.6. The code, therefore, exists over $GF(16)$ and its minimal polynomials are

$$\psi_1 = 1 + x + x^4$$
$$\psi_3 = 1 + x + x^2 + x^3 + x^4 \tag{8.18}$$

As the code exists in $GF(2^4)$ and every syndrome calculation structure needs m registers to store the remainder, the structure requires 4 flip-flops for the calculation of each syndrome. The calculation of the division remainder is performed in a way analogous to the encoding: as for the evaluation of the polynomial remainder, the logic matrix is represented by the adders below the registers. In particular for the first syndrome there is no adder, as the matrix for the evaluation of the first syndrome is the identity matrix.

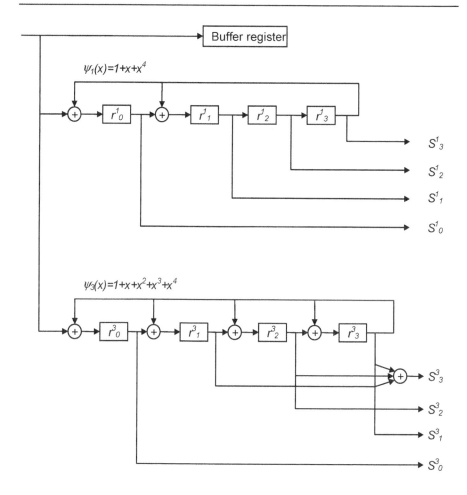

Fig. 8.6. Syndrome calculation for a code *BCH[15,11]*

For the second syndrome there is only one adder since only the third coefficient requires a sum. It is to this single adder that the contents of the registers containing the second, the third and the fourth coefficients of the polynomial remainder are input.

The considerations presented for the shortening operation during the encoding are also valid for the syndromes calculation. In fact also in this case the shortened message can ideally be filled in with 0s in the most significant bits, which are the first to enter the structure of calculation, up to the theoretical length.

Thus, also in this case the shortening does not modify the structure, whose latency is *n*.

A routine implementing the syndromes calculation is available in Appendix B (Sect. B.3).

8.3.2 Reed-Solomon

The syndromes calculation for Reed-Solomon codes is analogous to the one of BCH codes. Also in this case, as for the encoding, a suitable structure will group the flow of bits into symbols. The hardware and software calculations in this case coincide and it concerns the implementation of the Horner rule described in Eq. (8.12). However, the hardware structure for the calculation of the i-th syndrome, shown in Fig. 8.7, may also be seen as a pipelined polynomial division as it was for BCH codes.

If the structure is interpreted as a polynomial division, the process is analogous to the encoding but much simpler since the divisors, that is the minimal polynomials, are of first degree. Thus, there is only one feedback term, only one multiplication for α^i and only one register of m bits. Each input symbol of m bits is added to the output of the multiplication.

Note that, in this case, we do not need to evaluate the remainder in α^i as it is necessarily a polynomial of degree 0, i.e. a constant, since we are dealing with a first degree divisor.

Alternatively, the circuit may be interpreted as an implementation of the Horner rule, for which the incoming symbol is added to the content of the register before being multiplied by α^i and the result returns to the register.

The calculation finishes when all the n symbols enter the structure. There are $2t$ of these structures, one for each syndrome that has to be computed.

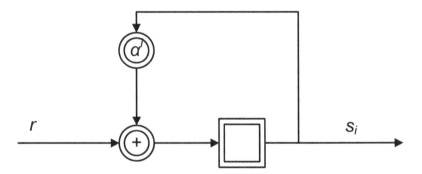

Fig. 8.7. Syndrome calculation for Reed-Solomon codes

The structure does not change also when considering shortened codes for the same reasons presented for BCH codes. If the shortened message is $n=2^s+2mt$ bit long the syndromes are produced in $[n/m]$ clock pulses where m is the number of bits forming a symbol.

8.4 Finding error locator polynomial

8.4.1 Binary BCH

The most complicated step in the decoding procedure of a cyclic code concerns the search for the error locator polynomial.

As already underlined in Chap. 2, the direct method is useful only for theoretical purposes and cannot be implemented in practical applications due to the fact that the computational complexity grows as the square of the correction capability.

The most used algebraic method to perform this step of the decoding is the Berlekamp-Massey algorithm.[4] The complexity of this algorithm grows in a linear way, enabling the construction of efficient decoders able to correct dozens of errors.

The philosophy of the Berlekamp algorithm consists in solving the set of Eqs. (2.64) in an iterative way for consecutive approximations.

At the i-th step of the algorithm we find a polynomial $\Lambda(x)$ whose coefficients solve the first i equations. Then, we test if $\Lambda(x)$ also solves the equation $i+1$; if not, we calculate the discrepancy term d so that $\Lambda(x)+d$ solves the first $i+1$ equations. After $2t$ iterations $\Lambda(x)$ is the error locator polynomial.

In the binary case it is possible to perform the Berlekamp algorithm in t iterations. The initial conditions are those given in Table 8.3.

Table 8.3. Initial condition for Berlekamp Massey algorithm

μ	$\Lambda^\mu(x)$	d_μ	l_μ	$2\mu{-}l_\mu$
$-1/2$	1	1	0	-1
0	1	S_1	0	0

The algorithm fills in the t lines of the table according to the followings steps.

- If $d_\mu{=}0$

$$\Lambda^{(\mu+1)}(x) = \Lambda^{(\mu)}(x)$$
$$l_{\mu+1} = l_\mu$$

(8.19)

- If $d_\mu{\neq}0$ then we look for another line ρ where $d_\rho{\neq}0$ and $2\rho{-}l_\rho$ is maximum.

$$\Lambda^{(\mu+1)}(x) = \Lambda^{(\mu)}(x) + d_\mu d_\rho^{-1} x^{2(\mu-\rho)} \Lambda^{(\rho)}(x)$$
$$l_{\mu+1} = \max\left(l_\mu, l_\rho + 2(\mu - \rho)\right)$$

(8.20)

- In any case $d_{\mu+1}$ is given by:

$$d_{\mu+1} = S_{2\mu+3} + \Lambda_1^{(\mu+1)} S_{2\mu+2} + \ldots + \Lambda_{l_{\mu+1}}^{(\mu+1)} S_{2\mu+3-l_{\mu+1}}$$

(8.21)

where $\Lambda_i^{(\mu+1)}$ represents the i-th coefficient of $\Lambda^{(\mu+1)}(x)$.

---

[4] E. R. Berlekamp, "Algebraic Coding Theory", *McGraw-Hill*, 1968.

- Increase μ and compute $2\mu - l_\mu$.
- After t steps we have $\Lambda^{(t)}(x)$: the error locator polynomial.

The best implementation for the algorithm is software, since we need to store all the lines of the table, to be able to make the choice later in case of discrepancy different from zero.

To effectively implement the Berlekamp algorithm in hardware we need to re-formulate the previous steps according to the flow diagram of Fig. 8.8.

First of all, we define the syndrome polynomial and we give the initial conditions as shown by Eq. (8.22).

$$1 + S = 1 + S_1 z + S_2 z^2 + \ldots + S_{2t-1} z^{2t-1}$$

$$\Lambda^{(0)}(z) = 1 \qquad\qquad d^{(0)} = 1 \tag{8.22}$$

At the i-th step we proceed as follows:

- if S_{2i+1} is unknown the algorithm is finished;
- otherwise we define $\Delta^{(2i)}$ the coefficient of z^{2i+1} in the product $(1+S(z))\Lambda^{(2i)}(z)$.

$$\Lambda^{(2i+2)}(z) = \Lambda^{(2i)}(z) + \Delta^{(2i)} \cdot d^{(2i)}(z) \cdot z \tag{8.23}$$

$$d^{(2i+2)}(z) = \begin{cases} z^2 d^{(2i)}(z) & \text{if } \Delta^{(2i)} = 0 \text{ or if } \deg \Lambda^{(2i)}(z) > i \\ \dfrac{z\Lambda^{(2i)}(z)}{\Delta^{(2i)}} & \text{if } \Delta^{(2i)} \neq 0 \text{ or if } \deg \Lambda^{(2i)}(z) \leq i \end{cases} \tag{8.24}$$

The polynomial $\Lambda^{(2t)}(z)$ is the error locator polynomial.

Note that, reformulated in this way, it is not necessary to store all the lines anymore. In Sect. B.6.1 an implementation of the Berlekamp-Massey algorithm is presented. The difficulty lies in the calculation of the inversion in Eq. (8.24). As explained in Chap. 6, the inversion in the Galois field is one of the most complicated operations to perform. One way to avoid the inversion is to use the flow diagram of Fig. 8.9.

Once defined the polynomial syndrome as above, the initial conditions are given as follows:

$$v^{(0)} = 1 \qquad k^{(0)} = 1 \quad \text{and} \quad \delta^{(2i)} = 1 \quad \text{if } i < 0 \tag{8.25}$$

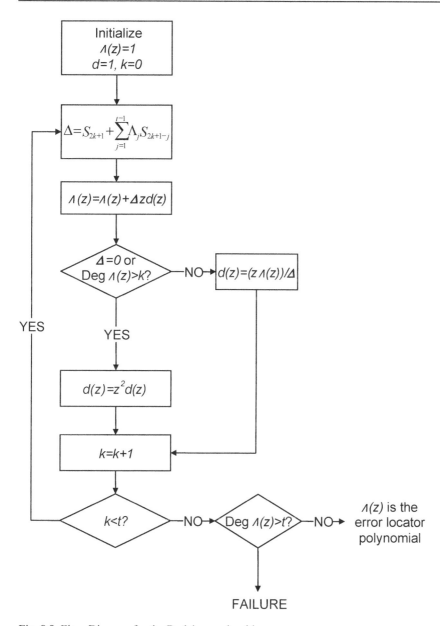

Fig. 8.8. Flow Diagram for the Berlekamp algorithm

We define $d^{(2i)}$ the coefficient of z^{2i+1} in the product $(1+S(z))v^{(2i)}(z)$.

- If S_{2i+1} is unknown the algorithm is finished;
- otherwise

$$v^{(2i+2)}(z) = \delta^{(2i-2)}v^{(2i)}(z) + d^{(2i)}k^{(2i)}(z) \cdot z \qquad (8.26)$$

$$k^{(2i+2)}(z) = \begin{cases} z^2 k^{(2i)}(z) & \text{if } d^{(2i)} = 0 \text{ or if } \deg v^{(2i)}(z) > i \\ z v^{(2i)}(z) & \text{if } d^{(2i)} \neq 0 \text{ or if } \deg v^{(2i)}(z) \leq i \end{cases} \qquad (8.27)$$

$$\delta^{(2i)} = \begin{cases} \delta^{(2i-2)} & \text{if } d^{(2i)} = 0 \text{ or if } \deg v^{(2i)}(z) > i \\ d^{(2i)} & \text{if } d^{(2i)} \neq 0 \text{ or if } \deg v^{(2i)}(z) \leq i \end{cases} \qquad (8.28)$$

Although this is not the error locator polynomial defined in Chap. 2, the roots of $v^{(2i)}(z)$ coincide with those of $\Lambda^{(2i)}(z)$.

In Sect. B.6.2 an implementation of the inversion-less Berlekamp-Massey algorithm is described.

In the following, advantages and disadvantages of Berlekamp algorithms with and without inversion are analyzed.

The degree of $\Lambda(z)$ and of $v(z)$ is the same but the null degree term of Λ is always 1, while the null degree term of v may be each field element. For this reason we need one more register to store the coefficients of $v(z)$. Besides, in the inversion-less algorithm it is necessary to calculate a quantity δ, so another register is needed to store it. The remaining operations are similar in both the cases except for the inversion in the classical Berlekamp-Massey algorithm and the extra multiplication for the modified one.

Considering that the inversion requires either a look-up table or a sequence of $2m-1$ multiplications, it is clear that this second method gains in terms of area or time.

Once the coefficients of the error locator polynomial are known, the errors are located by looking for the inverse of the roots.

Note that the degree of the error locator polynomial is t at the most. It represents the number of errors that most probably occurred. As already said above, the second and the third step of the whole decoding process are performed only in case errors occurred. Error detection is done at the end of the calculation of the syndromes, since having all the syndromes identically null is a necessary and sufficient condition to affirm that the read message is a codeword.

In case of shortened codes the procedure is exactly the same as the one described above, because the coefficients of the error locator polynomial depend exclusively on the syndromes calculated in the first step of the decoding process.

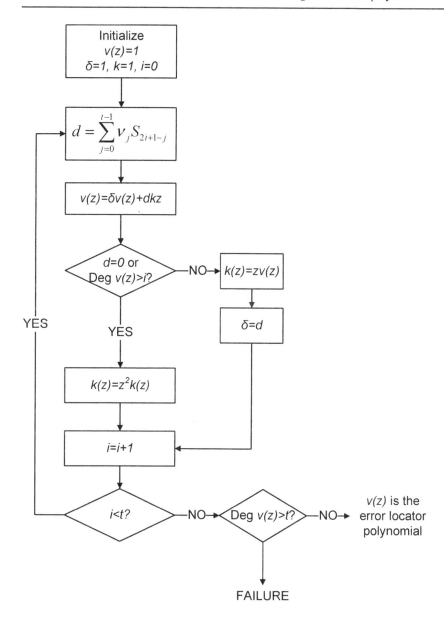

Fig. 8.9. Flow Diagram for the inversion-less Berlekamp algorithm

8.4.2 Reed-Solomon

Also for Reed-Solomon codes, one of the most efficient methods to perform the second step of the decoding process is the Berlekamp-Massey algorithm. For Reed-Solomon codes, however, the simplification of the binary case is not possible and the algorithm needs $2t$ steps.

Also for Reed-Solomon codes this step of the decoding procedure is performed only when errors have been detected, that is, when not all the syndromes are identically null.

The Berlekamp algorithm is now described.

1. Initialize the variables:

$$k = 0, \quad L = 0, \quad \Lambda^{(0)}(x) = 1, \quad T(x) = x \tag{8.29}$$

2. $k=k+1$. Calculate the discrepancy $d^{(k)}$:

$$d^{(k)} = S_k - \sum_{i=1}^{L} \Lambda_i^{(k-1)} S_{k-i} \tag{8.30}$$

3. If $d^{(k)}=0$ go to step 7.
4. The polynomial $\Lambda(x)$ is calculated as follows

$$\Lambda^{(k)} = \Lambda^{(k-1)}(x) - d^{(k)}T(x) \tag{8.31}$$

5. If $2L \geq k$ then go to step 7
6. If $L=k-L$ then $T(x)=\Lambda^{(k-1)}(x)/d^{(k)}$.
7. $T(x)=xT(x)$
8. If $k <2t$ go to step 2, otherwise $\Lambda(x)$ is the error locator polynomial.

The computational complexity is not very different from the one presented for BCH codes. The greatest difference lies in the fact that here $2t$ iterations instead of t are required.

Also for Reed-Solomon codes the best implementation of the Berlekamp-Massey algorithm is the software one. Again there is an inversion, but it is useless to search for an implementation avoiding it, since, as we will see later on, it will not be the last point in the decoding procedure where an inversion is required.

Besides, for Reed-Solomon codes also the Euclide division algorithm is efficient to perform this second step of the decoding process.

The Euclide algorithm is a fast method to find the greatest common divisor of a set of elements. The greatest common divisor of a set of elements can be expressed as a linear combination of the elements themselves. The algorithm operates on two elements (a,b) at the same time. Given the initial conditions $r_{-1}=a$, $r_0=b$, $s_{-1}=1$, $s_0=0$, $t_{-1}=0$, $t_0=1$, the algorithm proceeds according to the following recursive formulas:

$$r_i = r_{i-2} - q_i r_{i-1} \qquad \text{where } r_i < r_{i-1}$$
$$s_i = s_{i-2} - q_i s_{i-1} \tag{8.32}$$
$$t_i = t_{i-2} - q_i t_{i-1}$$

The algorithm finishes when $r_n=0$. The remainder r_{n-1} is the greatest common divisor between a and b. The recursion relations ensure that, at any given point in the algorithm, r_i can be expressed as $s_i a + t_i b$.

At the end, being d the greatest common divisor we have:

$$ua + vb = d \tag{8.33}$$

where u and v are coefficients produced by the algorithm.

In order to explain how the Euclide algorithm is used in the decoding process of a Reed-Solomon code, it is necessary to introduce two new polynomials.

Similarly to what described for BCH codes, the syndrome polynomial is defined as:

$$S(x) = S_{2t-1} x^{2t-1} + \ldots + S_1 x + S_0 \tag{8.34}$$

where the coefficients are the $2t$ syndromes calculated from the read word in the first step of the decoding routine.

As already said, Reed-Solomon codes work on symbols, so the location of the errors is not enough for the whole correction, but it is also necessary to know the magnitude of the errors.

Thus, we define the *error magnitude polynomial* as:

$$\Omega(x) = \Omega_{2v-1} x^{2v-1} + \ldots + \Omega_1 x + \Omega_0 \tag{8.35}$$

where v is the number of errors really occurred (it must be $v \le t$).

The syndrome polynomial, the error locator polynomial and the error magnitude polynomial are linked together by the Eq. (8.36) called *Key Equation*:

$$\Omega(x) = [S(x)\Lambda(x)] \bmod x^{2t} \tag{8.36}$$

The product of the polynomial $S(x)$ of degree $2t-1$ by the polynomial $\Lambda(x)$ of degree v is a polynomial of degree $2t+v-1$.

Equation (8.36) can then be rewritten as:

$$\Lambda(x) \cdot S(x) + F(x) \cdot x^{2t} = \Omega(x) \tag{8.37}$$

The known terms $S(x)$ and x^{2t} correspond to a and b of Eq. (8.33). The algorithm consists in dividing x^{2t} by $S(x)$ to produce a remainder. Then $S(x)$ becomes the dividend and the remainder calculated in the previous step becomes the new divisor in order to produce a new remainder. The process continues until the degree of the remainder is less than t. At the end, both the remainder $\Omega(x)$ and the multiplicative factor $\Lambda(x)$ are known.

The Euclide algorithm well suits both a software implementation and a hardware one due to its regular structure.

Also for Reed-Solomon codes, the roots of the error locator polynomial are the inverse of the error positions. Besides, also here the degree of the error locator polynomial establishes the number of errors that most probably have occurred.

Note that, since the calculation depends exclusively on the syndromes, shortened codes are implemented in the same way as described above.

8.5 Searching polynomial roots

8.5.1 Binary BCH

Once the error locator polynomial is known we have to find its roots, that is, the inverse of the error positions.

The algorithm used to search the roots, known as Chien algorithm, is a method based on trial and error. Substantially each field element is substituted in the error locator polynomial: if it satisfies the equation it is a root, otherwise the following element is tested. The inverse of the found root indicates an error location.

The error locator polynomial $\Lambda(x)$ of degree t at the most, for a code $BCH[n,k]$, is defined as:

$$\Lambda(x) = 1 + \Lambda_1 x + \dots + \Lambda_t x^t \qquad (8.38)$$

Verifying if a field element α^i satisfies the equation means verifying Eq. (8.39):

$$1 + \Lambda_1 \alpha^i + \dots + \Lambda_t \left(\alpha^i\right)^t = 0 \qquad (8.39)$$

If the equation is not satisfied the following element is considered, otherwise α^i is a root. In this case the inverse is an error position. Remembering the operations over the Galois field $GF(2^m)$ described in Chap. 6 we can affirm that the position 2^m-1-i is the erroneous one.

Using a software implementation, the substitution of n elements in the error locator polynomial of degree t requires nt multiplications and nt additions.

The hardware structure used to perform the third step of the decoding is represented in Fig. 8.10. The figure refers to a code $BCH[15,7]$ corrector of two errors.

The circuit requires t multipliers to multiply the content of the t registers by α, $\alpha^2,\dots, \alpha^t$. At the beginning the registers are loaded with the coefficients of the error locator polynomial $\Lambda_1, \Lambda_2,\dots, \Lambda_t$. The registers are shifted n times; at the i-th shift the registers contain $\Lambda_1 \alpha^i, \Lambda_2 \alpha^{2i},\dots, \Lambda_t \alpha^{ti}$ and the test described by Eq. (8.39) is performed.

If the test is satisfied, then 2^m-1-i is an error position, otherwise we shift the multipliers and we increase i. The search is complete in 2^m-1 clock pulses.

This structure becomes very onerous increasing the field's dimension and the correction capability of the code. If a fixed latency is not required, it is possible to save time depending on the real error positions. This time saving is based on the following observation: a polynomial of degree g with coefficients in a Galois field $GF(2^m)$ has g roots in the field at the most.

As already said above, at the end of the Berlekamp algorithm the degree of the polynomial $v \leq t$ indicates the number of errors that most probably occurred. As it will be explained in the next sections, this means that v errors occurred unless we have exceeded the correction capability of the code, so that the code erroneously recognize as v a number of errors higher than t. Nevertheless, the above observation is valid also in this case. The reason is that it is mathematically impossible for a polynomial of degree v to have a number of roots higher than v.

By exploiting this property it is possible to insert a counter in the structure so that whenever a root is found, its value is increased. When the number of roots already found is equivalent to the degree of the error locator polynomial, the search finishes, even if not all the field elements have been tested.

The effective advantage depends on the error locations inside the message. Suppose we have a code $BCH[2^m-1, 2^m-1-m]$ corrector of one error and that this error occurs in the last position, that is, in the position 2^m-1.

The Berlekamp algorithm gives out an error locator polynomial of first degree $1 + \Lambda_1 x$. The Chien machine begins with performing the test for $\alpha^0 = 1$, in order to verify if Eq. (8.40) is satisfied.

$$1 + \Lambda_1 = 0 \qquad (8.40)$$

Since we have supposed the last position as the erroneous one, the equation turns out to be satisfied and the root of the error locator polynomial is found. Due to the fact that we are dealing with a 1-degree polynomial (since the code corrects only one error), the Chien search is then terminated in a single clock pulse. In the worst case, i.e. the error in the first position, the latency is 2^m-1, that is, the same one of the classical search algorithm.

In case of shortened codes $[n, k]$ we do not need to substitute the 2^m-1 field elements.

As already said in Chap. 2, the codeword of a shortened code can be seen as a codeword of theoretical length containing a number of 0s equivalent to the number of shortened positions in the most significant bits.

Accordingly, a code $BCH[2^{m-1}+mt, 2^{m-1}]$ is equivalent to $BCH[2^m-1, 2^m-1-mt]$ where the last $q=2^m-1-2^{m-1}$ positions are identically null and therefore not considered in encoding and decoding operations. Obviously, these q positions must be errors-free, so that it is not useful to perform the Chien search algorithm for those elements whose corresponding error position is indeed one of those zeroes.

It is therefore a question of initializing the machine so that it explores only the field subset meaningful for the code that is being considered.

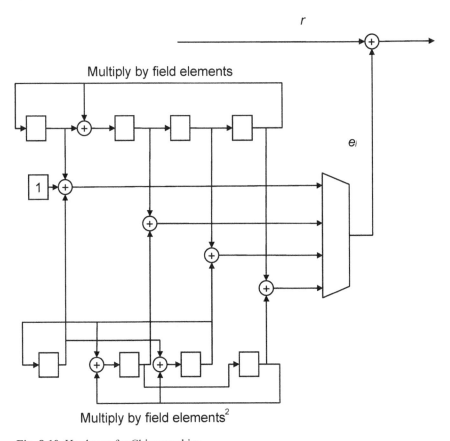

Fig. 8.10. Hardware for Chien machine

Accordingly, only the elements whose inverse is in a valid position of the message must be considered; that is, only the elements α^i, whose exponents i satisfy the Eq. (8.41), are tested.

$$0 < 2^m - 1 - i \le 2^{m-1} + mt \qquad (8.41)$$

Solving for i we get that the exponent of the initialization coefficient C is:

$$C = \alpha^{2^m - 1 - 2^{m-1} - mt} \qquad (8.42)$$

Thus, the structure described in Fig. 8.10 will not have the registers filled in with the coefficients $\Lambda_1, \Lambda_2, \ldots, \Lambda_t$ but with $C\Lambda_1, C\Lambda_2, \ldots, C\Lambda_t$. In Sect. B.7 an implementation for the Chien algorithm is described.

8.5.2 Reed-Solomon

Reed-Solomon codes require the same algorithm for the search of the roots of the error locator polynomial.

In this case, since we are operating on symbols rather than on bits, the operators are not binary anymore, but in the Galois field $GF(2^m)$.

Reed-Solomon codes integrate this step of the decoding process with the last step used for the calculation of the error magnitude, as it will be described in the next section.

8.6 Computing error magnitude

The last step of the decoding algorithm, corresponding to the search of the error magnitude, is performed only for non binary codes and will be here described in relation to Reed-Solomon codes.

The Euclide algorithm described in the preceding section enables us to solve Eq. (8.36), so that the coefficients of the error magnitude polynomial $\Omega(x)$ are known. In order to obtain the error magnitudes from $\Omega(x)$ an operation similar to the derivative is used. It is then necessary to introduce the definition of *formal derivative*, as the concept of derivative cannot be applied to finite fields, since we cannot apply the limits. However the formal derivative is computed like a derivative but it does not have the same interpretation. In other words, the formal derivative cannot be seen as a limit of a differentiation or as the tangent slope of a function.

Definition 8.6.1   Let $f(x)=f_0+f_1x+f_2x^2+...+f_nx^n$ be a polynomial with coefficients in $GF(q)$. The *formal derivative* $f'(x)$ is defined as:

$$f'(x) = f_1 + 2f_2x + ... + nf_nx^{n-1} \tag{8.43}$$

Given this definition, the Forney formula allows us to compute the magnitude of the errors.

The magnitude of the errors Y_j is given by:

$$Y_j = X_j \frac{\Omega(X_j^{-1})}{\Lambda'(X_j^{-1})} \tag{8.44}$$

where $\Lambda'(X_j^{-1})$ is the formal derivative of the error locator polynomial $\Lambda(x)$ evaluated in the roots found by the Chien algorithm.

Observe that this formula is valid only in case of errors, otherwise it makes no sense to find magnitudes.

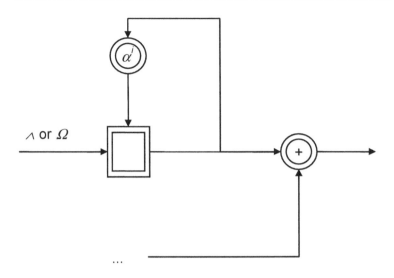

Fig. 8.11. Basic element for Chien/Forney algorithm

Also note that

$$\Lambda'\left(X_j^{-1}\right) = \Lambda_1 + \Lambda_3 X_j^{-2} + \Lambda_5 X_j^{-4} \tag{8.45}$$

Accordingly, $\Lambda'(X_j^{-1})$ is easily obtained from the error locator polynomial by setting the coefficients in even position at zero and dividing by (X_j^{-1}).

The hardware structure of the Forney algorithm is entirely analogous to the one of the Chien algorithm, whose basic element is represented in Fig. 8.11.

Since the basic element of the structure is analogous, we generally have a single hardware machine that contemporarily performs both the Chien and the Forney algorithms. Thus, the global structure of the third and fourth steps of the decoding process for a Reed-Solomon code is the one represented in Fig. 8.12. The circuit underlines the strong link between the Forney and the Chien algorithms, whose odd terms are directly used.

The crucial operation for the Forney algorithm is the inversion represented by the element "INV". As already seen in Chap. 6, performing an inversion in a Galois field is not a simple operation. In particular, given the hardware structure introduced above, the operation needs to be fast and efficient.

The use of a ROM meets both these requirements, even if it weights on the occupied area, particularly if we are using long codes. Finally, in case of shortened codes, the structure does not change since, as already said, the Forney algorithm gives a result only with the use of valid error positions.

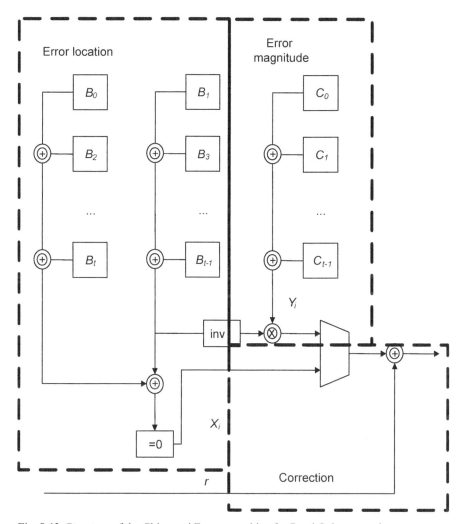

Fig. 8.12. Structure of the Chien and Forney machine for Reed-Solomon code

8.7 Decoding failures

BCH and Reed–Solomon codes have a good disclosing capability.

As described in Chap. 2, BCH and Reed-Solomon codes are not perfect codes: for this reason it is difficult that a codeword with more than t errors moves in the correction sphere of another codeword. The codewords of BCH and Reed-Solomon codes are well separated one from another and only a number of errors much greater than t could partially overlap their correction spheres.

This is the reason why, when more than t errors occur, the decoding process fails but erroneous corrections are not performed. It is therefore possible to use an error message showing that more than t errors have occurred.

Suppose we have a message containing more than t errors and see how the decoding proceeds.

At the exit of the syndromes calculation block it is not possible to detect if the correction capability has been exceeded; on the contrary, the calculation is completed with success by finding t or $2t$, apparently valid, syndromes.

The syndromes are transferred to the block that searches for the error locator polynomial. It has been said that the Berlekamp algorithm is a recursive algorithm that searches for the coefficients of the error locator polynomial using following approximations, by adding to the i-th iteration a discrepancy term d so that $\Lambda(x)+d$ solves the previous i equations. The discrepancy term is a monomial that is added to the error locator polynomial previously found. When the degree of the monomial to be added is greater than t, the correction capability of the code has been exceeded. In fact, it has to be remembered that the degree of the error locator polynomial is equal to the number of errors that most likely occurred. This number is reliable up to t; after t only the detection capability is reliable. In this case the decoding terminates with an error message.

Unfortunately it is not granted that, when the correction capability of the code is exceeded, this is what happens. On the contrary, most of the times this does not happen and the error locator polynomial apparently seems a valid one with a degree smaller than or equal to t (most of the times equal to t). Consequently, these coefficients, apparently valid, are loaded into the Chien machine.

When the correction capability of the code is exceeded, the Chien algorithm discloses it, since one of the following cases occurs:

- there are coincident roots;
- a sufficient number of roots is not found. Remember that a number of roots equal to the degree of the error locator polynomial has to be found;
- in case of shortened codes it can also happen that the shortened positions, those ones ideally filled in with 0s, are recognized as erroneous.

In practical implementations the first condition is never verifiable because, thanks to the described structure of the Chien machine, the same element is never tested more than once.

The second condition is the one that actually occurs in practice. At the end of the Chien algorithm we verify, through a comparator, if the number of roots found is equal to the degree of the error locator polynomial. If this condition is not satisfied, an error message shows that the correction capability of the code has been exceeded.

Finally, the third condition never occurs with the described Chien algorithm for shortened codes, because the use of an "initialization" constant avoids the testing of shortened positions.

However, remember that if the number of errors is much greater then the correction capability, the received message can be found in a correction sphere of another codeword: in this case the code might not be able to understand if the correction capability has been exceeded and might perform erroneous corrections.

8.8 BCH vs Reed-Solomon

The choice of the more suitable cyclic code for a particular application is not simple and it would be useful to know the distribution and the generation of the errors in details.

For this reason, it is not possible to say whether a Reed-Solomon code is superior or inferior to a binary BCH code. The greatest difference between the two codes is the correction basic unit, as BCH code corrects bits, while the Reed-Solomon code corrects symbols. Therefore, a BCH code corrects t bit errors, while the Reed-Solomon code corrects t symbols of s bit errors, which can be a variable number of bits: from t bits if errors occur in different symbols up to st bits in case errors occur consecutively. At a first glance this may seem an advantage, which however is not exploited if the error source is not of burst type (Sect. 1.5.3).

The choice of a code can also be based on the occupied area. In this case we have to note that both hardware implementation, if selected, and the number of parity bits contribute to increase area. Regarding this last aspect, fixing the same correction capability, Reed-Solomon codes require a greater number of parity bits, as shown in Table 8.4.

For what concerns the encoding implementation, the difference is surely that binary codes require binary operators (AND and XOR), while Reed-Solomon codes require operators in $GF(2^m)$ which have a greater computational complexity, as explained in Chap. 6.

Besides, since Reed-Solomon codes also require a greater number of parity bits, also the generator polynomial has a higher degree, and consequently the number of operators results to be higher. The remainders of the divisions must be stored into registers. Given the structure of the two codes, while BCH codes require the storage of bits, for Reed-Solomon codes it is necessary to store symbols.

Table 8.4. Number of parity bits required by Hamming inequality (minimum), BCH and RS codes

Data bits	Min $t=1$	BCH $t=1$	RS $t=1$	Min $t=2$	BCH $t=2$	RS $t=2$	Min $t=3$	BCH $t=3$	RS $t=3$	Min $t=4$	BCH $t=4$	RS $t=4$
2048	12	12	18	22	24	36	31	36	54	40	48	72
4096	13	13	18	24	26	36	34	39	54	44	52	72
8192	14	14	20	26	28	40	37	42	60	48	56	80
16384	15	15	22	28	30	44	40	45	66	52	60	88
32768	16	16	24	30	32	48	43	48	72	56	64	96

The hardware structure of the syndromes calculation involves the same considerations made for the encoding: the Reed-Solomon codes require a greater number of operators with a greater complexity. Besides, Reed-Solomon codes require the calculation of $2t$ syndromes, a number doubled if compared to binary BCH codes which exploit the binary property:

$$S_{2i} = S_i^2 \qquad\qquad\qquad (8.46)$$

Consequently also the inputs to the Berlekamp algorithm are doubled in comparison to a BCH code.

If the iterative Berlekamp algorithm is chosen, BCH codes require t iterations instead of the $2t$ required by Reed-Solomon codes.

The same considerations made for the encoding implementation apply to the Chien algorithm. Despite the fact that the number of operators is not actually greater, their complexity, as well as the registers for the storage, are more onerous in the case of Reed-Solomon codes compared to BCH ones.

Finally, note that Reed-Solomon codes require a further step in the decoding, represented by the Forney algorithm. The most onerous operation in this step is the inversion. As already said in Sect. 8.6, the operation is not simple and it requires an additional ROM in order to be efficient.

For what said so far, it is evident that if the choice criterion is the occupied area, BCH codes are by far superior to Reed-Solomon codes.

When the choice criterion becomes the latency then the situation changes and the BCH codes turn out to have inferior performances compared to Reed-Solomon codes. The difference lies in the fact that BCH codes are serial and process one bit at a time, while Reed-Solomon codes, operating on symbols, are defined with a parallel structure.

In Table 8.5 there is a comparison in terms of area and latency between BCH codes, Reed-Solomon codes using a ROM to perform the inversion and Reed-Solomon codes using the method of sequential inversion (Chap. 6). We have considered the case of correction of 4 bits or 4 symbols over 4096 bits.

In conclusion, Reed-Solomon codes generally have a greater correction capability for burst type errors, but they require a bigger area. On the contrary, BCH codes have a smaller area but a higher latency time.

The best compromise should probably be the decrease of BCH latency, also at the cost of increasing the area. In the next chapter we will explain how this can be obtained.

Table 8.5. Number of clock pulses needed for computation

		Encoder	Syndrome	Berlekamp	Chien/Forney	Tot
Average latency	RS (seq.)	528	528	528/50	5280/50	645
	RS (ROM)	528	528	528/59	528/50	550
	BCH	4148	4148	1/50	4148/50	4231
Maximum latency	RS (seq.)	528	528	528	5280	6336
	RS (ROM)	528	528	528	528	1584
	BCH	4148	4148	1	4148	8297
Area (Eq. Gates)	RS (seq.)	1120	1120	1120	1560	3800
	RS (ROM)	485	485	485	458	1428
	BCH	1012	4975	485	916	6371

Bibliography

S. R. Blackburn, W. G. Chambers, "Some Remarks on an Algorithm of Fitzpatrick" in *IEEE Transactions on Information Theory*, Vol. 42, July 1996.

H. O. Burton, "Inversionless Decoding of Binary BCH Codes" in *IEEE Transactions on Information Theory*, Vol. 17, July 1971.

H.-C. Chang, C. B. Shung, "New Serial Architecture for the Berlekamp-Massey Algorithm" in *IEEE Transactions on Communications*, Vol. 47, April 1999.

H.-C. Chang, C. B. Shung, "A Reed-Solomon Product-Code (RS-PC) Decoder Chip for DVD Applications" in *IEEE Journal of Solid-State Circuits*, Vol. 36, February 2001.

C. L. Chen, "High-Speed Decoding of BCH Codes" in *IEEE Transactions on Information Theory*, Vol. 27, March 1981.

C. K. P. Clarke, "Reed-Solomon Error Correction" in *British Broadcasting Corporation, R&D White Paper*, 2002.

S. V. Fedorenko, P. V. Trifonov, "Finding Roots of Polynomials Over Finite Fields" in *IEEE Transactions on Communications*, Vol. 50, November 2002.

G. L. Feng, K. K. Tzeng, "A Generalized Euclidean Algorithm for Multisequence shift-Register Synthesis" in *IEEE Transactions on Information Theory*, Vol. 35, May 1989.

G. L. Feng, K. K. Tzeng, "A Generalization of the Berlekamp-Massey Algorithm for Multisequence Shift-Register Synthesis with Applications to Decoding Cyclic Codes" in *IEEE Transactions on Information Theory*, Vol. 37, September 1991.

P. Fitzpatrick, "On the Key Equation" in *IEEE Transactions on Information Theory*, Vol. 41, September 1995.

P. Fitzpatrick, S. M. Jennings, "Comparison of Two Algorithms for Decoding BCH Codes" in *ISIT*, June 1997.

D. Forney, "On Decoding BCH Codes" in *IEEE Transactions Letters on Information Theory*, 1965.

V. Guruswami, M. Sudan, "Improved Decoding of Reed-Solomon and Algebraic-Geometry Codes" in *IEEE Transactions on Information Theory*, Vol. 46, September 1999.

M. J. Jarchi, S. K. Sridharan, "Architecture for Multi-Symbol Encoding and Decoding" in *US Patent 2003/0106013 A1*, June 2003.

Y. Kijima, "Apparatus for Decoding BCH Codes" in *US Patent 5,208,815*, May 1993.

C. H. Kraft, "BCH Error-Location Polynomial Decoder" in *US Patent 5,343,481*, August 1994.

J. N. Laneman, C. E. W. Sundberg, "Reed-Solomon Decoding Algorithms for Digital Audio Broadcasting in the AM Band" in *IEEE Transactions on Broadcasting*, Vol. 47, June 2001.

H. Lee, "High-Speed VLSI Architecture for Parallel Reed-Solomon Decoder" in *IEEE Transactions on Very Large Scale Integration (VLSI) Systems*, Vol. 11, April 2003.

Y. S. Lin, Y. S. Kang, "Error Correction Circuit for BCH Codewords" in *US Patent 5,416,786*, May 1995.

M. Lunelli, R. Micheloni, R. Ravasio, A. Marelli, "Method for Performing Error Corrections of Digital Information Codified as a Symbol Sequence", *US Patent US20050050434*, 2005.

J. Massey, "Shift-Register Synthesis and BCH Decoding" in *IEEE Transactions on Information Theory*, Vol. 15, January 1969.

R. Micheloni, R. Ravasio, A. Marelli, "Method and System for Correcting Low Latency Errors in Read and Write Non Volatile Memories, Particularly of the Flash Type", *US Patent US20060010363*, 2006.

R. Micheloni, R. Ravasio, A. Marelli, "Method and System for Correcting Errors in Electronic Memory Devices", *US Patent US2006005109*, 2006.

R. Micheloni, R. Ravasio, A. Marelli, "Reading Method of a Memory Device With Embedded Error Correcting Code and Memory Device With Embedded Error Correcting Code", *European Patent EPA06425141.6*, 2006.

S. Mizrachi, D. Stopler, "Efficient Method for Fast Decoding of BCH Binary Codes" in *US Patent 2003/0159103 A1*, August 2003.

A. Raghupathy, K. J. Ray Liu, "Low Power/High Speed Design of Reed Solomon Decoder" in *IEEE International Symposium on Circuits and Systems*, June 1997.

K. J. Ray Liu, A. Y. Wu, A. Raghupathy, J. Chen, "Algorithm-Based Low-Power and High-Performance Multimedia Signal Processing" in *Proceedings of the IEEE*, Vol. 86, June 1998.

I. S. Reed, M. T. Shih, T. K. Truong, "VLSI Design of Inverse-Free Berlekamp-Massey Algorithm" in *IEEE Proceedings*, Vol. 138, September 1991.

I. S. Reed, M. T. Shih, T. K. Truong, "VLSI Design of Inverse-Free Berlekamp-Massey Algorithm" in *Proc. Inst. Elect. Eng. Pt. E*, Vol. 138, September 1991.

J. Sankaran, D. Hoyle, "Error Correction Structures and Methods" in *US Patent 2002/0002693 A1*, January 2002.

D. V. Sarwate, N. R. Shanbhag, "Very High-Speed Reed-Solomon Decoders" in *ISIT2000*, June 2000.

D. V. Sarwate, N. R. Shanbhag, "High-Speed Architectures for Reed-Solomon Decoders" in *IEEE Transactions on Very Large Scale Integration (VLSI) Systems*, Vol. 9, October 2001.

D. D. Sullivan, "A Branching Control Circuit for Berlekamp's BCH Decoding Algorithm" in *IEEE Transactions on Information Theory*, September 1972.

Z. Szwaja, "On Step-by-Step Decoding of the BCH Binary Codes" in *IEEE Transactions on Information Theory*, April 1967.

T. K. Truong, J. H. Jeng, I. S. Reed, "Fast Algorithm for Computing the Roots of Error Locator Polynomials up to Degree 11 in Reed-Solomon Decoders" in *IEEE Transactions on Communications*, Vol. 49, May 2001.

W. B. Weeks, W. D. Little, V. K. Bhargava, T. A. Gulliver, "Comparison of the Motorola DSP56000 and the Texas Instruments TMS320C25 Digital Signal Processors for Implementing a Four Error Correcting (127,99) BCH Error Control Code Decoder" in *IEEE Pacific Rim Conference on Communications, Computers and Signal Processing*, June 1989.

W. Wilhelm, "A New Scalable VLSI Architecture for Reed-Solomon Decoders" in *IEEE Journal of Solid-State Circuits*, Vol. 34, March 1999.

J. K. Wolf, "ECC Performance of Interleaved RS Codes with Burst Errors" in *IEEE Transactions on Magnetics*, Vol. 34, January 1998.

T. Zhang, K. K. Parhi, "On the High-Speed VLSI Implementation of Errors-and Erasures Correcting Reed-Solomon Decoders" in *GVLVLSI*, April 2002.

9 BCH hardware implementation in NAND Flash memories

9.1 Introduction

The continuous pressure of the market on memory manufacturers to reduce costs and to increase capacitance, pushes the semiconductor technology towards more and more extreme geometries. In NAND memories with a critical dimension (CD) of 32 nm, the cell is programmed by injecting some hundred electrons into its floating gate (FG), the loss even of one single electron a month will involve, in the space of 10 years, the loss of information. Today, there are also NAND memories which, discriminating more levels of the charge present in the floating gate, are able to store more information bits in each cell. The native error probability provided by these new memories is such to require the correction of more than one error bit on a packet of data.

We will now see how a BCH coding can effectively offer a simple solution to get the required error probability. We will see how it is possible to integrate a BCH code into a system and in particular how to realize it by integrating it into a NAND MLC memory.

9.2 Scaling of a ECC for MLC memories

For single level memories, where each memory cell contains one single information bit, the scaling of a ECC simply requires us to know the error probability desired BER_{out} and the native error probability BER_{in}. The correction code to be applied must allow us to fill this gap. To correctly face the scaling of a correction code for multilevel memories, we need to understand the meaning of native error probability in the presence of a coding of more information bits in the same cell.

In 4-level memories two information bits are memorized in the same memory cell. These four levels are generically represented with the probability distribution of the threshold voltages of cells when these are programmed with equal probability on the four levels (Fig. 9.1).

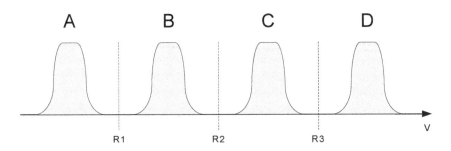

Fig. 9.1. Representation of the probability distribution of the V_t of cells when programmed with equal probability on the four distributions

In principle, we would like these four distributions to be always separate but, due to disturbs and retention, described in Chap. 5, we get an overlapping area on the tails of the distributions. We will call $\psi_{A,RX}$ the area subtended by the distribution "A" starting from the reference 'RX' towards the extremes of the distribution (Fig. 9.2).

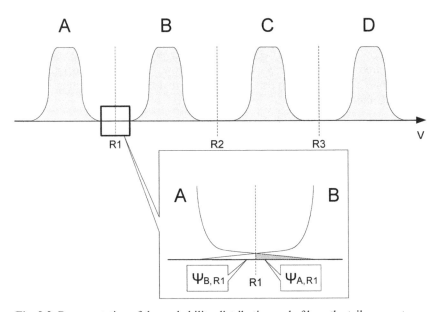

Fig. 9.2. Representation of the probability distribution and of how the tails generate an error

The four distributions must be encoded. The criteria to allocate the four codes are mainly linked to the requisites of reliability and performance of the read and program operations. The Gray code and the Decimal codes are commonly used (Chap. 6, Sect. 6.1).

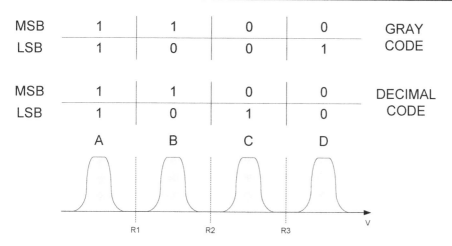

| MSB | 1 | 1 | 0 | 0 | GRAY |
| LSB | 1 | 0 | 0 | 1 | CODE |

| MSB | 1 | 1 | 0 | 0 | DECIMAL |
| LSB | 1 | 0 | 1 | 0 | CODE |

Fig. 9.3. Representation of the probability distribution of the V_t of cells when programmed with equal probability on the four distributions and, coding of the two information bits contained in each single cell with the Gray and Decimal codes

The two bits, which we will call LSB and MSB, may belong to the same page or to different pages. It is necessary to clarify the effect that the coding and the allocation of the bit on the page produces on the error probability. Consider the four possible combinations in the hypothesis of a 2112-Byte page. With the Gray code the MSB bit is obtained with a single read with respect to the reference R2 (Fig. 9.3) and we get an error probability which is approximately equal to:

$$P_{MSB} = \psi_{B,R2} + \psi_{C,R2},$$ (9.1)

where we have left out the contributions deriving from the farest distributions from R2, the A and the D ones. By considering all the error terms equal to ψ we get:

$$P_{MSB} = 2\psi$$ (9.2)

With the same hypotheses, the LSB bit has an error probability equal to:

$$P_{LSB} = \psi_{A,R1} + \psi_{B,R1} + \psi_{C,R3} + \psi_{D,R3} = 4\psi.$$ (9.3)

If the LSB and MSB bits belonging to a cell are all allocated on one page, we have that the error probability of the page is:

$$P_{EPage}(\psi) = 1 - (1 - P_{LSB})^{8448}(1 - P_{MSB})^{8448}$$
$$= 1 - (1 - 4\psi)^{8446}(1 - 2\psi)^{8448}$$ (9.4)

The *BER* for these pages is:

$$BER = 3\psi \tag{9.5}$$

If the LSB and MSB bits belonging to a cell are allocated on different pages we have that the error probabilities of the pages containing LSB bits and MSB bits differ and they are respectively:

$$P_{EPageMSB} = 1 - \left(1 - P_{MSB}\right)^{16896} = 1 - \left(1 - 2\psi\right)^{16896} \tag{9.6}$$

$$BER = 2\psi \tag{9.7}$$

$$P_{EPageLSB} = 1 - \left(1 - P_{LSB}\right)^{16896} = 1 - \left(1 - 4\psi\right)^{16896} \tag{9.8}$$

$$BER = 4\psi \tag{9.9}$$

Likewise, for the decimal encoding we respectively get:

$$P_{MSB} = 2\psi \tag{9.10}$$

$$P_{LSB} = 6\psi \tag{9.11}$$

If the MSB and LSB bits belonging to a cell are all allocated on one page, we have that the error probability of the page is:

$$P_{EPage} = 1 - \left(1 - P_{LSB}\right)^{8448}\left(1 - P_{MSB}\right)^{8448} = 1 - \left(1 - 6\psi\right)^{8448}\left(1 - 2\psi\right)^{8448} \tag{9.12}$$

$$BER = 4\psi \tag{9.13}$$

If the LSB and MSB bits belonging to a cell are allocated on different pages, we have that the error probabilities of the pages containing LSB bits and MSB bits differ and they are respectively:

$$P_{EPageMLS} = 1 - \left(1 - P_{MSB}\right)^{16896} = 1 - \left(1 - 2\psi\right)^{16896} \tag{9.14}$$

$$BER_{MSB} = 2\psi \tag{9.15}$$

$$P_{EPageLSB} = 1 - \left(1 - P_{LSB}\right)^{16896} = 1 - \left(1 - 6\psi\right)^{16896} \tag{9.16}$$

$$BER_{LSB} = 6\psi \qquad\qquad (9.17)$$

The smallest error probability is obtained by means of a Gray coding with the LSB and MSB bits of each cell used on the same page (Fig. 9.4). In applications which maintain the LSB and MSB bits of each cell on separate pages, the analysis of the BER must be studied and treated separately for the LSB pages and for the MSB ones.

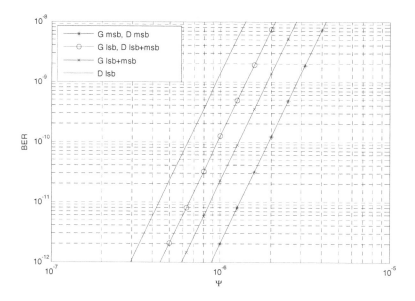

Fig. 9.4. Representation of the BER_{out} as a function of the BER_{in} in the hypothesis of correction of 5 bits with 2112-Byte pages in the range 10^{-6}–10^{-5}

Having to realize and size an error correction system for NAND memories we need to understand immediately that this must be sized for the worst pages. It would be a serious mistake to use a mediate BER_{in} for scaling.

Since it is important to consider the correct BER_{in}, for the scaling of the system, it is good to highlight the other causes of diversity of the BER_{in} on the pages of NAND memories.

- The pages of a NAND structure are usually divided in even/odd. These two categories of pages do not necessarily introduce the same BER_{in}. FG coupling, program algorithms and specification limits can induce a difference in the BER_{in} between odd and even pages.

- The block of a NAND memory is typically constituted by 32 WLs. The edge WLs generally suffer more for disturbs and stress induced by program and read operations. The pages allocated on these edge WLs can induce a BER_{in} more elevated than the other WLs.

A simple technique that allows us to uniform the BER_{in} consists in the scrambling of the bits with different error probability in the unit which will constitute the data block to correct. This for example occurs in the Gray and decimal coding with LSB and MSB bits on the same page. This is not always tolerated by the system but, where applicable, it has great efficiency. For example, we may consider using data units larger than the single page, e.g. couples of LSB and MSB pages, or consider units which involve the whole WLs to uniform the BER_{in} of LSB/MSB and even/odd. Where the system allows it, we could go as far as to treat the whole NAND block as the unit subject to correction. A key point for a correct scaling of the system correction capability, besides the BER_{out} required by the system, is therefore the BER_{in} expressed in its components (even/odd, LSB/MSB and so on) when not uniform for all the matrix cells.

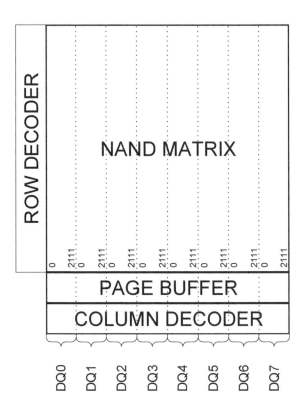

Fig. 9.5. Representation of how BLs are organized in a NAND matrix to produce DQs

The possible correlation between the errors in somehow adjacent bits is another important parameter to know and it provides some indications on the most suitable correction code to use. In particular a Reed-Solomon code could benefit from a correlation between the bits belonging to the same symbol. Actually, as described in Chap. 5 on the reliability and the causes of an onset of error in Flash memories, the errors are predominantly associated to native defects or defects arisen in the oxide of the single cells and there are not many elements of correlation. Then observe that the pages are alternated on Even and Odd BLs and the BLs which give a certain output bit (DQ) are usually all grouped in contiguous matrix portions. In this way, we get a sort of scrambling (Fig. 9.5) during the data input and the data output so that even in the hypothesis of physical correlation of the defects at stack or BL level, where the cells are organized, this scrambling makes any attempt of correlation complex.

9.3 The system

In this chapter we will assume to have a system formed by a control unit which provides and receives data from a NAND MLC memory with page composed by 2112 Bytes. To guarantee the required BER_{out} we will assume that the system requires the correction of 5 errors on the whole page and that it is able to signal the presence of 6 or more errors with elevated probability. The code used must be a $BCH[16896, 16821]$. For such coding we first need to obtain the following elements:
the primitive polynomial is

$$\phi_1 = (1100.0000.0000.0001) = 1 + x + x^{15} \tag{9.18}$$

The minimal polynomials for syndrome computation are

$$\psi_1 = (1100.0000.0000.0001) = 1 + x + x^{15} \tag{9.19}$$

$$\psi_3 = (1100.0100.0010.0001) = 1 + x + x^5 + x^{10} + x^{15} \tag{9.20}$$

$$\psi_5 = (1101.0000.0000.1001) = 1 + x + x^3 + x^{12} + x^{15} \tag{9.21}$$

$$\psi_7 = (1101.0101.0101.0101)$$
$$= 1 + x + x^3 + x^5 + x^7 + x^9 + x^{11} + x^{13} + x^{15} \tag{9.22}$$

$$\psi_9 = (1110.1100.0010.0001) = 1 + x + x^2 + x^4 + x^5 + x^{10} + x^{15} \tag{9.23}$$

the generator polynomial of $BCH[2^{15}-1, 2^{15}-76]$

$$g = \psi_1 \cdot \psi_3 \cdot \psi_5 \cdot \psi_7 \cdot \psi_9 \tag{9.24}$$

which, once it has been worked out, gives:

$$g = (1110.0000.\,0110.1111.\,0011.0100.\,1010.0100.\,0011.0011.\,1110.$$
$$1010.1110.\,0110.1011.\,1001.0011.\,1111.1101\,) = 1 + x + x^2 + x^{10}$$
$$+ x^{11} + x^{13} + x^{14} + x^{15} + x^{16} + x^{19} + x^{20} + x^{22} + x^{25} + x^{27} + x^{30}$$
$$+ x^{35} + x^{36} + x^{39} + x^{40} + x^{41} + x^{42} + x^{43} + x^{45} + x^{47} + x^{49} + x^{50} \tag{9.25}$$
$$+ x^{51} + x^{54} + x^{55} + x^{57} + x^{59} + x^{60} + x^{61} + x^{64} + x^{67} + x^{68} + x^{69}$$
$$+ x^{70} + x^{71} + x^{72} + x^{73} + x^{74} + x^{75}$$

The method to obtain these polynomials is described in Chap. 8. The number of parity bits is equal to the degree of the generator polynomial, 75 bits. The code efficiency is equal to:

$$\frac{DataField}{CodeWord} = \frac{CodeWord - ParityBit}{CodeWord} = \frac{2112 \cdot 8 - 75}{2112 \cdot 8} = 0.9956 \tag{9.26}$$

The functional architecture of a BCH correction code, already described in the previous chapters, can easily be adapted to memories and described as in Fig. 9.6.

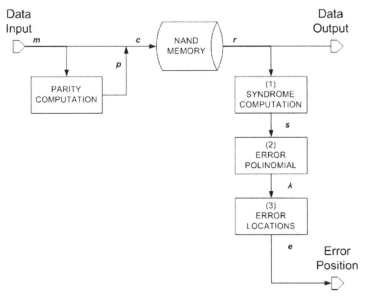

Fig. 9.6. Functional representation of the coding and decoding of a BCH code

Let's now briefly go through the flow of operations.

- In writing operations, the datum is sent to the memory and to the "Parity Machine" block for parity computation. Data (d) and parity (p) constitute the codeword (c) which is written in the memory.
- In reading operations, the message read by the memory (r) is contemporarily sent to the external controller and to the "Syndrome Machine" block, for the computation of the syndromes (s). The controller saves the part of the code associated to the message.

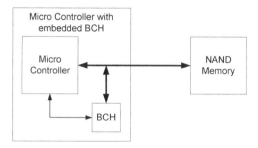

Fig. 9.7. Representation of a system with BCH ECC embedded in the microcontroller

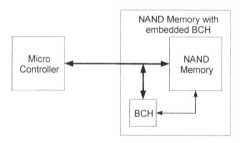

Fig. 9.8. Representation of a system with BCH ECC embedded in a NAND Memory

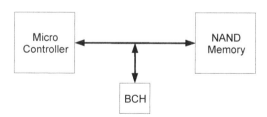

Fig. 9.9. Representation of a system with BCH ECC stand-alone

- If all the syndromes are null, the message is recognized as valid, we do not proceed with any other operations and the message can immediately be made available.
- If one or more syndromes are different from the null vector, the message contains one or more errors.
- Depending on the value of the syndromes, the "Berlekamp" block calculates the error locator polynomial and the number of errors occurred.
- Depending on the result of the elaboration of the "Berlekamp" block, the "Chien Machine", for the search of the roots of the error locator polynomial, is initialized and started.
- The Chien Machine executes an ordered scanning of all the possible error positions and actually signals the error positions until it has located all the errors indicated by the Berlekamp machine.
- The list of error positions is sent to the controller which accurately and quickly operates the right inversions to correct the received code.
- Once the corrections are executed the data will be made available.

Alternatively, the BCH machine could directly perform the corrections on the message but this would require a re-transmission of the whole message. In general, to increase the performance and reduce the system latency, it is convenient, for this phase of correction, to be directly performed by the controller.

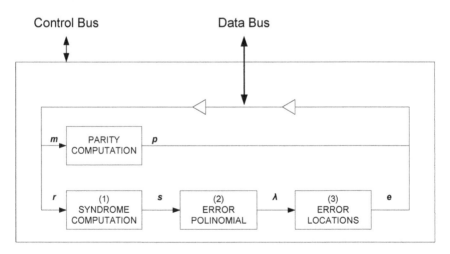

Fig. 9.10. Representation of the machines forming the BCH correction block

There exist many possible implementations of a system constituted by controller, BCH processor and NAND memory. The BCH processor can be embedded into the controller, into the NAND memory or stand-alone as a separate processor (Figs. 9.7 to 9.9). In all the cases the BCH processor (Fig. 9.10) receives the data to encode or to decode on one single parallel bus (Data Bus) with words of 8, 16 bits, while the control of the flow of the operations is ideally submitted to a Control Bus. The BCH coding and decoding operations are described in the vast majority of the cases as a process which operates in a serial way on a sequence of bits. Hence the difficulty to adopt it in a system that operates on words. There are at least two possible approaches for the realization of a BCH processor. The first one simply consists in the serialization of the data from/to the NAND memory so as to bring them back to a sequence of bits. The second approach consists in defining machines able to elaborate the received data in a parallel way. Before carrying on with the examination of a particular architecture, we will now analyze how to realize the single machines forming the code *BCH[16896, 16821]*.

9.4 Parity computation

The organization of the system by words induces us to fix the dimension of the message at an integer number of bytes. The bytes 0,1,...N constitute the message *m*, and are numbered depending on the sequence in which they are written or read by the memory (Fig. 9.11). To correctly operate the BCH coding and decoding, it is useful to associate the higher degree of the polynomial *m(x)* to the most meaningful bit of the byte(0) of the message, and the degree 0 to the least meaningful bit of the last byte of message. The bits containing the parity are encoded likewise. Since the parity uses 75 bits, the five least meaningful bits in the tenth parity byte are not used and we will assume them of null value. The code word *c(x)* is formed by the parity bits in the least meaningful positions followed by the message word. The code we get is usually a shortened code, all the missing positions to get a complete codeword can be assumed null. We will see that these stratagems help us to perform the coding and decoding operations more easily.

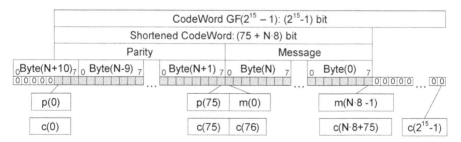

Fig. 9.11. Layout of the bytes constituting the message, the parity and the BCH code

As described by Eqs. (9.27), the parity computation is obtained as a remainder of the division of the message *m(x)* by the generator polynomial *g(x)*.

$$\frac{m(x) \cdot x^{n-k}}{g(x)} = q(x) + \frac{p(x)}{g(x)}$$

$$m(x) \cdot x^{n-k} + p(x) = q(x) \cdot g(x) \tag{9.27}$$

$$c(x) = m(x) \cdot x^{n-k} + p(x)$$

The implementation of a machine for the calculation of the parity in a system which operates by words can be carried out through the typical sequential architecture described in Chap. 8. This configuration requires the BCH processor to operate with a frequency synchronous with the data flowing and with a frequency with multiplicity equal to the number of bits present in the data word. The circuit and the sub-circuit in Fig. 9.12, illustrate a possible implementation for the parity computation, in the hypothesis of a system with 8-bit words. The serial parallel circuit (PtoS) converts the sequence of words in transit towards the memory (M) into a sequence of bits. Such sequence is applied to the division G (Fig. 9.13) which calculates its remainder.

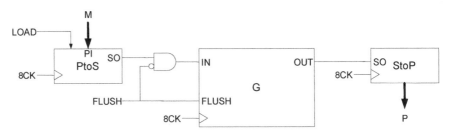

Fig. 9.12. Block representation of the machine for the parity computation by means of conversion of the flow of bytes of the message M into a sequential flow of bits m allocated to the serial divisor G which operates with a frequency depending on the dimension of the word. The parity bits which are taken as remainder of the division are converted into bytes again

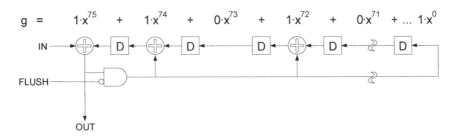

Fig. 9.13. Circuital representation of the polynomial divisor. The FLUSH signal, when active and with IN at "0", forces the circuit to operate as a shift register

By activating the FLUSH signal which forces the divisor input at '0' and forces it to operate as a shift register, the 75 parity bits are extracted by the control system through the parallel serial converter (StoP) as 10 bytes, where the last parity byte automatically finishes with 5 bits at 0.

If the BCH processor has a working frequency unable to bring the data flow back to a sequence of bits, it is necessary to use a parallel processing structure. The serial structure of the parity computation machine can be turned into a parallel machine following some simple steps (Fig. 9.14). The iterative operation carried out by the serial divisor to elaborate a word, can be unrolled. In this way, the following state of the registers is calculated on the basis of all the operations needed to elaborate a whole message word.

Sequential implementation

$$g = 1 \cdot x^{75} + 1 \cdot x^{74} + 0 \cdot x^{73} + 1 \cdot x^{72} + 0 \cdot x^{71} + \ldots 1 \cdot x^0$$

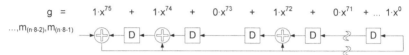

Unrolled implementation

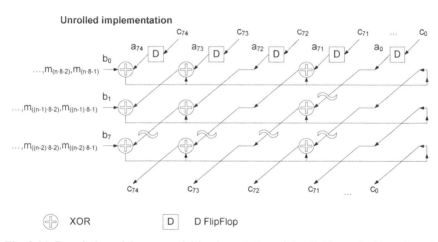

| ⊕ | XOR | | | D | D FlipFlop |

Fig. 9.14. Description of the sequential implementation of the divider and of how it could be unrolled to get a parallel implementation

The direct synthesis of the unrolled implementation of the divisor risks causing some complications to the synthesis tools due to the elevated number of inputs of the combinatorial logic and the depth of the paths.

The whole combinatorial part of the unrolled implementation can be expressed by describing each of its outputs as a function of the previous state a_w and of its inputs b_k.

$$c_j = \sum_{k=0}^{7} d_{kj} \cdot b_k + \sum_{w=0}^{74} l_{wj} \cdot a_w \ ;$$

$$d_{kj}, l_{wj} \in \{0,1\} \quad \forall \ j \in \{0,...,74\}$$

(9.28)

or in matrix form

$$C = (B|A) \cdot (D|L)^T$$

(9.29)

where

$$(B|A) = \{b_0, b_1, b_2, b_3, \ldots b_6, b_7, a_0, a_1, a_2, \ldots, a_{73}, a_{74}\}$$

(9.30)

$$(D|L) = \begin{cases} \begin{bmatrix} d_{0,0} & d_{0,1} & \cdots & d_{0,7} & l_{0,0} & l_{0,1} & \cdots & l_{0,74} \\ d_{1,0} & d_{1,1} & \cdots & d_{1,7} & l_{1,0} & l_{1,1} & \cdots & l_{1,74} \\ \cdots & \cdots & \cdots & \cdots & \cdots & \cdots & \cdots & \cdots \\ d_{73,0} & d_{73,1} & \cdots & d_{73,7} & l_{73,0} & l_{73,1} & \cdots & l_{73,74} \\ d_{74,0} & d_{74,1} & \cdots & d_{74,7} & l_{74,0} & l_{74,1} & \cdots & l_{74,74} \end{bmatrix} \end{cases}$$

(9.31)

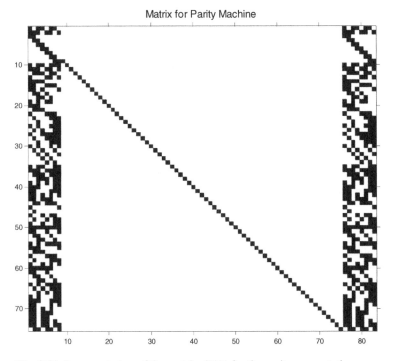

Fig. 9.15. Representation of the matrix *(D|L)* for the parity computation

In Appendix D (Sect. D.1), we can see a simple MatLab routine which generates such binary matrix with all the simplifications deriving when operating in *GF*. This matrix is graphically represented in Fig. 9.15, where the unitary coefficients are marked. In this form it is quite simple to get the direct conversion of the matrix into a VHDL code, as illustrated in Fig. 9.16. The code obtained is efficient in simulation and the synthesis tools can easily optimize the common terms and prepare the arithmetic operators in pyramidal form to get the best performances, as represented in Fig. 9.17.

```
architecture rtl of Parity is
begin
c(0) <= b(0) XOR b(2) XOR b(3) XOR b(5) XOR b(6) XOR
              a(67) XOR a(69) XOR a(70) XOR a(72) XOR a(73);
c(1) <= b(0) XOR b(1) XOR b(2) XOR b(4) XOR b(5) XOR b(7) XOR
              a(67) XOR a(68) XOR a(69) XOR a(71) XOR a(72)
  XOR a(74);
c(2) <= b(0) XOR b(1) XOR a(67) XOR a(68);
c(3) <= b(1) XOR b(2) XOR a(68) XOR a(69);
c(4) <= b(2) XOR b(2) XOR a(69) XOR a(70);
...
c(74) <= b(1) XOR b(2) XOR b(4) XOR b(5) XOR b(7) XOR
              a(66) XOR a(68) XOR a(69) XOR a(71) XOR a(72)
  XOR a(74);
end rtl;
```

Fig. 9.16. VHDL representation of the combinatorial net of the parity computation machine obtained as a simple implementation of the matrix representation

Short path implementation

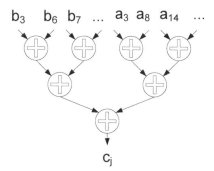

Fig. 9.17. Reduction in pyramidal form of the parity computation operators

The parity computation machine is composed by a bank of registers and a combinatorial net (RC) as sketched in Fig. 9.18. At the end of the input of the data message, the parity bits provided from the message by the system controller remain in the machine registers.

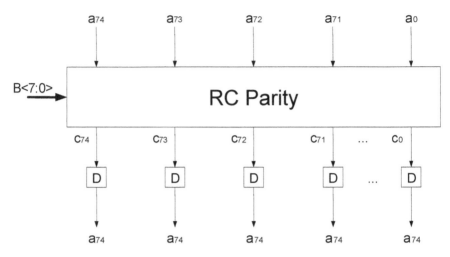

Fig. 9.18. Representation of the parallel machine for the parity computation

9.5 Syndrome computation

As described in Chap. 8, the syndrome computation can be carried out in two different ways. The first method obtains the syndromes as evaluation of the message $r(x)$ in α powers. By considering the received message as the sum of the original message and an error vector, we have:

$$r(x) = m(x) + e(x)$$
$$S = r \cdot H^T = (m + e) \cdot H^T$$

(9.32)

For the way the matrix H^T is built we have:

$$S = (S_1, S_3, S_5, S_7, S_9) = \left(r(\alpha), r(\alpha^3), r(\alpha^5), r(\alpha^7), r(\alpha^9)\right)$$
$$S_i = r(\alpha^i)$$

(9.33)

The syndromes can be rewritten as:

$$S_i = r(\alpha^i) = r_{n-1}(\alpha^i)^{n-1} + r_{n-2}(\alpha^i)^{n-2} + \ldots + r_1\alpha^i + r_0$$
$$= \left(\ldots\left(\left(r_{n-1}\alpha^i + r_{n-2}\right)\alpha^i + r_{n-3}\right)\alpha^i + \ldots + r_1\right)\alpha^i + r_0$$

(9.34)

which represents the Horner rule, where the syndromes can be obtained through an iterative operation of multiplication by α and sum of a message bit r. The second method uses the division of the received message $r(x)$ by the minimal polynomials $\Psi_i(x)$ evaluated in α. In fact:

$$r(x) = Q_i(x) \cdot \psi_i(x) + b_i(x) \tag{9.35}$$

where

$$b_i = remainder\left(\frac{r(x)}{\psi_i(x)}\right) \tag{9.36}$$

Since

$$S_i = r\left(\alpha^i\right)$$
$$\psi_i(\alpha^i) = 0 \tag{9.37}$$

we get that:

$$S_i = r\left(\alpha^i\right) = b_i(\alpha^i) \tag{9.38}$$

And therefore each syndrome is given by the evaluation in α_i of the remainder of the division $b_i(x)$. We will consider the system implementation using both these two methods. The system requires the computation of 5 syndromes of 15 bits (S_1, S_3, S_5, S_7, S_9). The computation of S_1 equal to $r(\alpha)$ with the Horner rule can be obtained with the following sequential machine (Fig. 9.19).

Sequential implementation

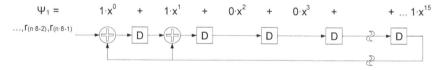

$$\Psi_1 = \quad 1 \cdot x^0 \quad + \quad 1 \cdot x^1 \quad + \quad 0 \cdot x^2 \quad + \quad 0 \cdot x^3 \quad + \quad \quad + \; \dots \; 1 \cdot x^{15}$$

$\dots, \Gamma_{(n \cdot 8-2)}, \Gamma_{(n \cdot 8-1)}$

Fig. 9.19. Representation of the machine for the computation of the syndrome S_1 as $r(\alpha)$

At each cycle, this machine executes the multiplication of the content of the register by α and the sum of a message bit r_i in the least meaningful position. The machines for the evaluation of $r(\alpha^3)$, $r(\alpha^5)$, $r(\alpha^7)$ and $r(\alpha^9)$ are all analogous and they simply require the use of a multiplier by constants α^3, α^5, α^7, α^9 respectively. Also for the syndrome computation we are interested in a parallel implementation; the unrolling technique applied to the parity computation can be used in a similar way. The resulting system for the syndrome computation is represented in Fig. 9.21.

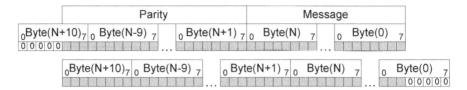

Fig. 9.20. The sequence constituting the message is read with the same alignment originated during the write operations, alignment on the Byte(0). For syndromes computation, it is convenient to shift the whole message to align it on the last byte, Byte(N+10)

The message $r(x)$, read by the memory, is addressed to the five syndrome computation machines. We have to keep in mind how the message $c(x)$ and therefore $r(x)$ had been assembled in the coding phase. The last message byte Byte(N+10) contains only three meaningful bits (Fig. 9.20). Since the syndrome computation machines have parallelism equal to the length of the word (e.g. 8 or 16 bits), it is convenient to shift the whole structure of the read message, to get an alignment at the end of the frame. In this way, the bits at 0 in the first word do not modify the result of the syndrome calculation and, once the last byte is received, the result of the syndromes is immediately available in the computation machine registers.

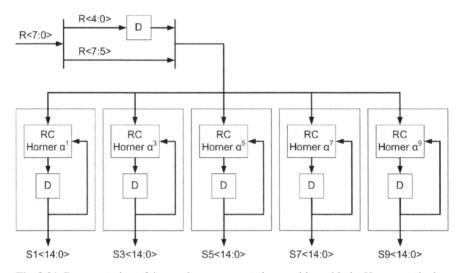

Fig. 9.21. Representation of the syndrome computation machine with the Horner method

For each of the combinatorial nets "RC Horner α^x", the outputs c_j can be expressed as a combination of the preceding state a_w and the inputs b_k:

$$c_j^i = \sum_{k=0}^{7} d_{kj}^i \cdot b_k + \sum_{w=0}^{14} l_{wj}^i \cdot a_w \; ; \tag{9.39}$$

$$d_{kj}, l_{wj} \in \{0,1\} \quad \forall \; j \in \{0,...,14\}$$

In matrix form:

$$C_j^i = D^i \times B + L^i \times A = \left(D^i \mid L^i\right) \times \left(B \mid A\right)^T \tag{9.40}$$

The matrices for the Horner combinatorial nets *(D|L)*, can be simply calculated (Appendix D) and are represented in Fig. 9.22.

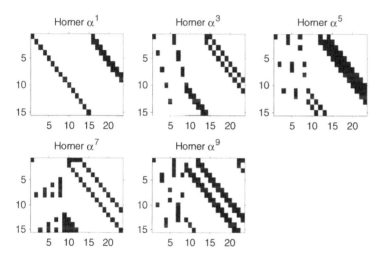

Fig. 9.22. Representation of the matrices for the syndrome computation with the Horner method

The second method uses the division of the received message *r(x)* for the minimal polynomials $\Psi_i(x)$ evaluated in α^i. The system deriving from it is represented in Fig. 9.23.

As for the implementation through the Horner rule, also in this case it is necessary to provide a shift of the received message to align it with the end of the message. The machine is constituted by five divisors which divide the message *r(x)* by the respective five primitive polynomials *(Ψ₁(x), Ψ₃(x), Ψ₅(x), Ψ₇(x), Ψ₉(x))*. The bits at zero inserted at the beginning do not alter the result of the divisions which, in this way, end aligned with the last received byte.

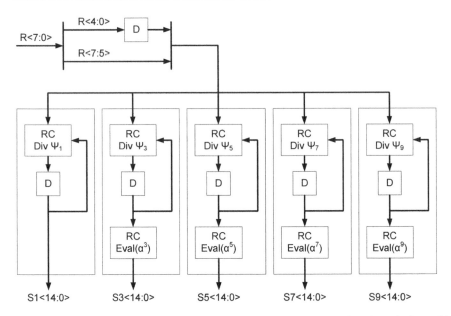

Fig. 9.23. Representation of the machine for the syndrome computation through the residual method

When the message is completely received, the five registers of the five machines contain the remainders of the divisions.

$$b_i = remainder\left(\frac{r(x)}{\psi_i(x)}\right)$$

$$i = \{1,3,5,7,9\}$$

(9.41)

The structure of the divisors is identical to the one used for the parity computation, the dimensions and the matrices (Fig. 9.24) which must be used for the generation of the logic nets "Div Ψ_x" are different (Appendix D).

To get the five syndromes S_1, S_3, S_5, S_7, S_9, these remainders have to be evaluated in α^1, α^3, α^5, α^7, α^9 respectively. As underlined in Chap. 8, the result of this evaluation can be expressed for each syndrome j as:

$$s_i^j = \sum_{k=0}^{14} d_{i,k}^j \cdot r_k^j \; ;$$

$$d_{k,j} \in \{0,1\} \quad \forall\, i \in \{0,...,14\}$$

(9.42)

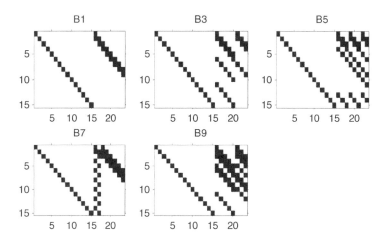

Fig. 9.24. Representation of the matrices B_1, B_3, B_5, B_7 and B_9 used for the generation of the logic nets Div Ψ_1, Div Ψ_3, Div Ψ_5, Div Ψ_7 and Div Ψ_9

In matrix form:

$$S_j = A_j \cdot R_j \qquad (9.43)$$

The matrices A_j (Fig. 9.25) can be easily calculated and are given in Appendix D.

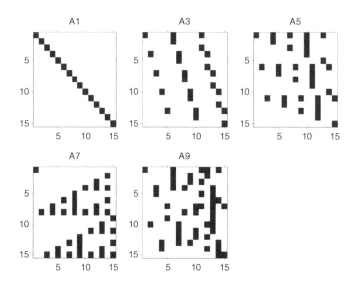

Fig. 9.25. Representation of the matrices A_1, A_3, A_5, A_7, A_9

The synthesis of the combinatorial net is carried out by direct implementation of Eqs. (9.42), as already shown for the combinatorial nets, which execute the divisions. A counter is also inserted into the syndrome computation machines to keep a trace of the number of bytes which form the message and whose value will then be used by the Chien machine for the determination of the error position. Besides, a signal (ErrorDetected), active if all the syndromes are not null, signals the presence of one or more error to be corrected in the received message.

9.6 Berlekamp machine

The Berlekamp algorithm is an algorithm commonly used for the calculation of the coefficients of the error locator polynomial. The execution of this algorithm can be performed through a common processor, if it is possible to have a complete table of the elements of the field to quickly perform multiplications in $GF(2^{15})$. Since such table usually has unsuitable dimensions for small controllers of memories and of memory cards, the possible approaches are two. The first approach requires us to add to the core of the controller a hardware for the execution of the operations of sum and multiplication in $GF(2^{15})$, an opportune set of instructions for the management of the new operations and a reduced number of registers to manipulate all the intermediate variables of the Berlekamp recursive algorithm. The architecture deriving from it is represented in Fig. 9.26.

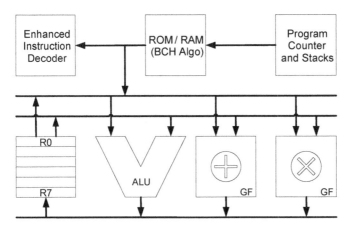

Fig. 9.26. Representation of the architecture of the controller modified in order to include the essential operators to execute the Berlekamp algorithm

The adder $GF(2^{15})$ is reduced to a XOR operation between two 15-bit terms and does not need further consideration. We need to pay some attention to the area and the consumption of the multiplier in $GF(2^{15})$. In Fig. 9.27 there is a possible implementation of the multiplier.

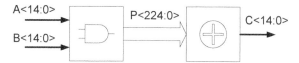

Fig. 9.27. Representation of the multiplier in $GF(2^{15})$

The terms a and b first produce the vector p such that:

$$p_{ij} = a_i \wedge b_j$$
$$\forall i, j \in \{0,1,...,14\} \tag{9.44}$$

Starting from this vector p, the terms c of the product in $GF(2^{15})$ can be directly obtained as a combination of the terms of p.

$$c_i = \sum_{k=0}^{224} m_{k,i} \cdot p_k;$$
$$m_{k,i} \in \{0,1\} \quad \forall \, i \in \{0,...,14\} \tag{9.45}$$

The matrix M is available in Appendix D with a simple function to obtain it. In this form the multiplier presents the minimum number of XOR operators and the consequent pyramidal structure provides good performances and low consumption. We do not give the details of a generic execution of the Berlekamp algorithm here, since they are given in Chap. 8, and we simply remind that after t complete iterations we get the error locator polynomial coefficients, the number of errors occurred or a signal of the detection of 6 or more errors which cannot be managed by the correction machine. The execution of the Berlekamp algorithm has a "remarkable" implementation, particularly efficient if the number of corrections is smaller or equal to 5. The origin and the execution of this implementation are now briefly described.

The closed solutions of the Berlekamp-Massey algorithm start from the first description of the iterative Berlekamp-Massey algorithm described in Chap. 8. The iterative algorithm can be executed symbolically at each row in the table. In other words, it is possible to fill the rows of the table using symbols for field quantities instead of actual quantities derived from received vectors. This kind of computation leads to a set of formulas for the polynomial coefficients instead of a computational algorithm. The decoding procedure evaluates these formulas rather than iteratively filling the table for each read message.

In the following we will see an example of how this table can be filled up for the five-error case (Table 9.1).

Table 9.1. Example of a closed solution for a branch of the Berlekamp tree

M	$\Lambda^\mu(x)$	d_μ		l_μ	$2\mu{-}l_\mu$
$-1/2$	1	1		0	-1
0	1	S_1		0	0
1	$1+ S_1 x$	$S_1^{3}+ S_3$	$(=0)$	1	1
2	$1+ S_1 x$	$S_1^{5}+ S_5$	$(=0)$	1	3
3	$1+ S_1 x$	$S_1 S_1^{3}+ S_7$	$(=0)$	1	5
4	$1+ S_1 x$				

In this example all the discrepancies found, except for the first one, are zero. In this case we have found the error locator polynomial for the single error case, that is when the following inequalities apply:

$$S_1 \neq 0$$
$$A = S_3 + S_1^{3} = 0$$
$$B = S_5 + S_1^{5} = 0 \qquad\qquad (9.46)$$
$$N = S_7 + S_1 S_3^{2} = 0$$

This case represents only a leaf of the whole tree.

When operating symbolically, there are two possible ways to fill each row, one with a null discrepancy and one with a nonzero discrepancy. Since both are possible, the symbolic computation has to consider both the possibilities. In this way, each row of the table must be replaced by a branch of a binary tree where it is possible to choose a different way depending on the current value of the discrepancy d. At each node, intermediate polynomials and other values are stored temporarily so that the next level can be computed; these values can be discarded once the level is reached.

The final result is a decision tree composed of 2^t branches where odd syndromes' components are used in equations over $GF(2^m)$ to decide the path to be followed from the root to the final branch. At every decision node the discrepancy introduced in Chap. 8 is computed, while each branch represents a set of formulas for the error locator polynomial coefficients for a particular read message. Vectors that cannot be corrected by the code may result in polynomials with degree greater than t or in polynomials that have no roots in the field (Chap. 10). This is the case of a detectable, but uncorrectable error pattern.

In Fig. 9.28 the tree for the five errors case is represented.

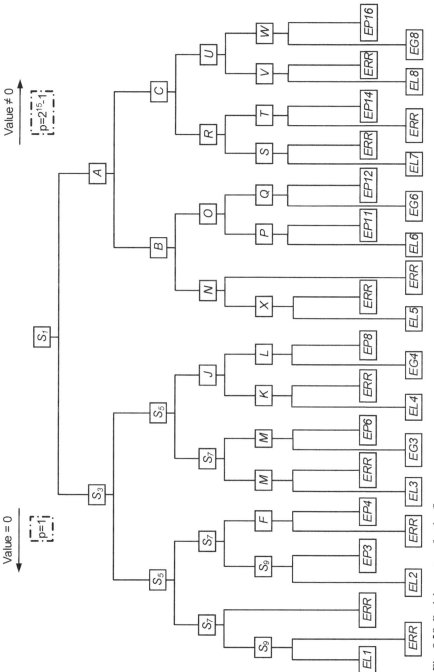

Fig. 9.28. Berlekamp tree for the five-error case

The variables represented in the tree are described by the following set of equations:

$$A = S_3 + S_1^3$$

$$B = S_5 + S_1^5$$

$$C = S_5 + S_1^5 + AS_1^{-1}S_3$$

$$F = S_9 + S_5^{-1}S_7^2$$

$$J = S_7 + S_5^2 S_3^{-1}$$

$$K = S_9 + S_5 S_3^{-1} S_7 + S_3^3$$

$$L = S_9 + S_5 S_3^{-1} S_7 + S_3^3 + JS_3^{-1} S_5$$

$$M = S_9 + S_3^3$$

$$N = S_7 + S_1 S_3^2$$

$$O = S_7 + S_1 S_3^2 + BS_1^{-1}S_3$$

$$P = S_9 + S_1^9 + BS_1^{-1}S_5$$

$$Q = S_9 + S_1^9 + OB^{-1}\left(S_7 + S_1 S_3^2\right) + BS_1^{-1}S_5$$

$$R = S_7 + S_1 S_3^2 + AS_1^{-1}S_5$$

$$S = S_9 + S_1^9 + AS_1^{-1}S_7$$

$$T = S_9 + S_1^9 + AS_1^{-1}S_7 + RA^{-1}\left(S_5 + S_1^5\right)$$

$$U = S_7 + S_1 S_3^2 + \left(AS_1^{-1} + CA^{-1}\right)S_5 + CA^{-1}S_1^5$$

$$V = S_9 + S_1^9 + \left(AS_1^{-1} + CA^{-1}\right)S_7 + CA^{-1}S_1 S_3^2$$

$$W = S_9 + S_1^9 + \left(AS_1^{-1} + CA^{-1} + UC^{-1}\right)S_7$$
$$+ \left(CA^{-1} + UC^{-1}\right)S_1 S_3^2 + UC^{-1}AS_1^{-1}S_5$$

$$X = S_9 + S_1^9$$

(9.47)

On the other hand, all the possible branches are represented by the following set of equations:

$$EL1 = 1$$
$$EL2 = 1 + S_5 x^5$$
$$EL3 = 1 + S_3 x^3$$
$$EL4 = 1 + S_5 S_3^{-1} x^2 + S_3 x^3$$
$$EL5 = 1 + S_1 x$$
$$EL6 = 1 + S_1 x + BS_1^{-1} x^4$$
$$EL7 = 1 + S_1 x + AS_1^{-1} x^2$$
$$EL8 = 1 + S_1 x + \left(CA^{-1} + AS_1^{-1}\right) x^2 + CA^{-1} S_1 x^3$$
$$EG2 = 1 + S_7 S_5^{-1} x^2 + S_5 x^5$$
$$EG3 = 1 + S_3 x^3 + S_7 S_3^{-1} x^4$$
$$EG4 = 1 + S_5 S_3^{-1} x^2 + S_3 x^3 + JS_3^{-1} x^4$$
$$EG6 = 1 + S_1 x + OB^{-1} x^2 + OB^{-1} S_1 x^3 + BS_1^{-1} x^4$$
$$EG7 = 1 + S_1 x + AS_1^{-1} x^2 + RA^{-1} x^4 + RA^{-1} S_1 x^5$$
$$EG8 = 1 + S_1 x + \left(AS_1^{-1} + CA^{-1} + UC^{-1}\right) x^2 \qquad (9.48)$$
$$\qquad + \left(CA^{-1} + UC^{-1}\right) S_1 x^3 + UC^{-1} AS_1^{-1} x^4$$
$$EP3 = 1 + S_9 S_5^{-1} x^4 + S_5 x^5$$
$$EP4 = 1 + S_7 S_5^{-1} x^2 + FS_5^{-1} x^4 + S_5 x^5$$
$$EP6 = 1 + MS_7^{-1} x^2 + S_3 x^3 + S_7 S_3^{-1} x^4 + MS_7^{-1} S_3 x^5$$
$$EP8 = 1 + \left(S_5 S_3^{-1} + LJ^{-1}\right) x^2 + S_3 x^3 + \left(JS_3^{-1} + LJ^{-1} S_5 S_3^{-1}\right) x^4$$
$$\qquad + LJ^{-1} S_3 x^5$$
$$EP11 = 1 + S_1 x + \left(BS_1^{-1} + PB^{-1}\right) x^4 + PB^{-1} S_1 x^5$$
$$EP12 = 1 + S_1 x + OB^{-1} x^2 + OB^{-1} S_1 x^3 + \left(BS_1^{-1} + QB^{-1}\right) x^4 + QB^{-1} S_1 x^5$$
$$EP14 = 1 + S_1 x + \left(AS_1^{-1} + TR^{-1}\right) x^2 + TR^{-1} S_1 x^3$$
$$\qquad + \left(RA^{-1} + TR^{-1} AS_1^{-1}\right) x^4 + RA^{-1} S_1 x^5$$
$$EP16 = 1 + S_1 x + \left(AS_1^{-1} + CA^{-1} + UC^{-1} + WU^{-1}\right) x^2$$
$$\qquad + \left(CA^{-1} + UC^{-1} + WU^{-1}\right) S_1 x^3 + WU^{-1} CA^{-1} S_1 x^5$$
$$\qquad + \left[UC^{-1} AS_1^{-1} + WU^{-1} \left(AS_1^{-1} + CA^{-1}\right)\right] x^4$$

Observe that the further right in the tree the computation goes, the more complicated the formulas become. This is equivalent to saying that there are some error patterns that finish the calculation very quickly and others which take a longer time. Generally, the quickest ones are those with fewer errors, while the worst case is the rightmost path from top to bottom.

The general decoding procedure, using the Berlekamp tree, can be summarized by the following steps:

1. Enter the tree at the top to get the first node.
2. At any node, compute the variable represented in the formulas of Eq. (9.47) and decide to go left or right in the tree to the next lower node. Go left if the variable is zero and go right if it is not zero.
3. If the current node is not a leaf, go back to step 2.
4. Choose the correct error locator represented in Eq. (9.48).

The process requires $t-1$ decisions. The needed formulas require only odd syndromes because we are dealing with binary BCH codes. Observe that some paths lead to an ERR square. This is when the degree of the error locator polynomial is greater than t and an uncorrectable condition is detected.

It should be noted that even though some of the equations seem very complicated, there are many terms that can be computed high in the tree and saved for later use. Besides, these computations proceed fairly fast as the sum is an Exclusive-OR, while the multiplication and especially the inverse is computed as described in Chap. 6.

The overall computational complexity for the procedure is the same as the one we have in the iterative algorithm, that is t^2, but it grows linearly with the growing of the number of errors. In particular it is possible to observe that the number of inversions, that is the most consuming operation, increases of one unit as the degree of the polynomial increases of one unit. Even if the computational complexity is the same, the number of computations required by the Berlekamp tree is smaller than in the iterative method. This is particularly true, because there are a lot of quantities that can be computed at the top and used later on without recalculating them.

Anyway, the process is computationally efficient up to about $t=5$. Beyond that, the tree gets too large and the iterative method can be more efficient. This is due to the fact that the formulas start to become quite long and complicated.

The main advantage of the tree solution over the iterative solution is that both the decision variables at each branch and the polynomial coefficients, are represented by formulas which can be evaluated in a straightforward manner. On the other hand the iterative algorithm must be followed exactly at each step to obtain the correct polynomial. Some common mistakes can be made in the implementation of the iterative method. While the closed solutions for the Berlekamp tree can prove correct, it is very difficult to prove that a given implementation of the iterative method behaves correctly for all the read messages.

The Berlekamp tree can be implemented via software or via hardware.

If the chosen implementation is the software one there will be some branches that require a few operations and some others that require a lot of operations. Then there will be some branches that are followed very often and some others that are hardly ever followed. Starting from this observation it is possible to cut the branches that are never followed, ensuring a gain in the number of instructions required by the microcontroller to implement the entire tree.

In order to choose the right branches to cut a lot of simulations have to be made. Looking at Fig. 9.28 observe that in case 3 errors occur, the error locator polynomial can be *EL3*, *EL4* and *EL8*, i.e. all the third-degree polynomials. Starting from a read message of length n, it is possible to generate 10^6 error patterns of three errors and see how many times the *EL3*, *EL4* and *EL8* branches are followed. In the specific case of $n=16896$ this simulation shows that the branch *EL3* is never followed. This means that the branch *EL3* is followed with a probability less than $p=10^{-6}$; since this probability is the target that a memory has to guarantee this branch can be cut from the "usual Berlekamp tree". This procedure can be repeated also for the *4* or *5* errors case, so that some branches can be eliminated from the tree.

The final tree is represented in Fig. 9.29, where the variables and the polynomials are the ones represented in Eq. (9.47) and Eq. (9.48).

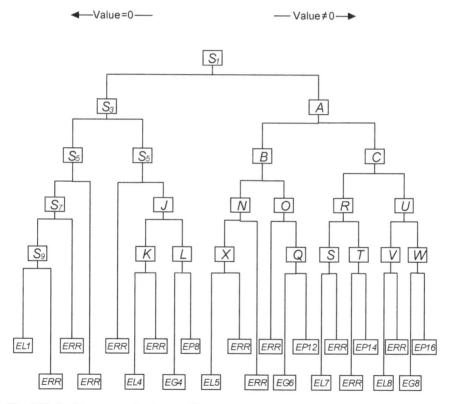

Fig. 9.29. Berlekamp tree for the specific case

Finally, observe that the branches to be cut depend on the specific BCH code chosen and this tree does not represent the general case. Every different BCH code, in particular every different length n can lead to different branches to be cut.

In Appendix D an example of the custom Berlekamp tree implementation is described.

The software approach for the realization of the Berlekamp algorithm presents the drawback of a sequential implementation and therefore of a large number of operations to perform. Although the number of operations to execute grows in a linear way, above a certain limit the small card and memory controller risks to introduce an excessive latency in the decoding procedure. The world of telecommunications and DVD decoders has identified extremely fast and efficient structures from which to draw when necessary. The state of the art is represented by the HW implementations of the Berlekamp inversion-less algorithm[1] (Fig. 9.30).

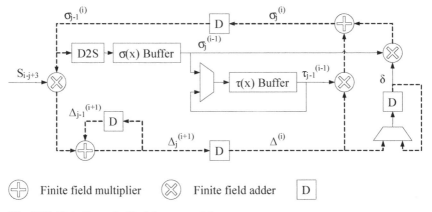

 Finite field multiplier   Finite field adder   D

Fig. 9.30. Data flow of a Berlekamp machine

As in all number crunching architectures the single feedback loop operations are directly instantiated. The system is composed by operators and registers (data path) and by a control system generally governed by a FSM (Finite State Machine) or by a core VLIW (Very Long Instruction Word) able to correctly perform the operations.

9.7 The Chien machine

Once the coefficients of the error locator polynomial have been calculated through the Berlekamp machine, the Chien machine searches all its roots. The inverses of such roots indicate the error positions.

$$\Lambda(x) = 1 + \Lambda_1 x + \ldots + \Lambda_t x^t \tag{9.49}$$

To determine the roots of the polynomial, the Chien machine orderly evaluates $\Lambda(x)$ in all the elements of the field α^0, α^1, α^2, α^3,... α^N. For each element i of the field for which the polynomial is null, the corresponding position $(2^{15} - 1 - i)$ is an error position. A possible implementation of the Chien machine is represented in Fig. 9.31.

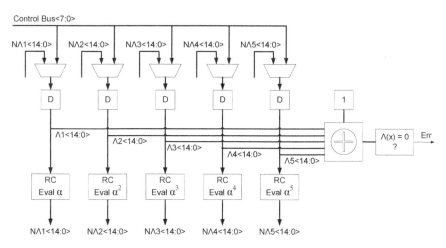

Fig. 9.31. Chien machine: sequential implementation

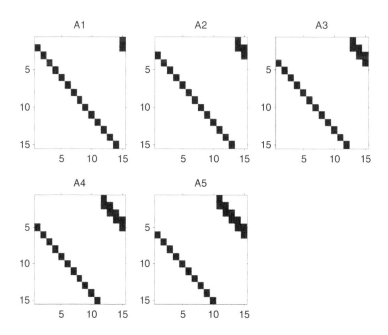

Fig. 9.32. Representation of the matrices for the multiplication by α^1, α^2, α^3, α^4 and α^5

At first the registers D are loaded with the values of the error locator polynomial coefficients, Λ_1, Λ_2, Λ_3, Λ_4 and Λ_5. The first evaluation of the polynomial is therefore performed for $x = \alpha^0$. If the sum of all the terms is null, α^0 is a root of $\Lambda(x)$ and we have an error bit in the position $2^{15}-1$. Beginning from the present state of the polynomial Λ_1, Λ_2, Λ_3, Λ_4 and Λ_5, we calculate the value of the polynomial in the following state $N\Lambda_1$, $N\Lambda_2$, $N\Lambda_3$, $N\Lambda_4$, $N\Lambda_5$ multiplying by the constants α^1, α^2, α^3, α^4 and α^5 respectively.

$$N\Lambda_i = A_i \cdot \Lambda_i$$
$$\forall\, i \in \{1,2,3,4,5\}$$
(9.50)

The matrices A (Appendix D) are represented in Fig. 9.32.

The Chien machine, thus realized, searches the roots one at a time. In this phase the system is not executing any operation, neither read nor write, towards the memory but, it is waiting for the signal of the error positions. Such operation, carried out for all the bits of the message, results to be a very onerous operation in terms of time. Also in this case it is possible to pass to a parallel realization. Unlike the parity and the syndromes computation machines, which have to operate with a parallelism equal to the input data parallelism, the Chien machine does not have particular limits other than the complexity, the area occupied and the consumption. Figure 9.33 shows the representation of a possible implementation of the Chien machine with an 8-bit parallelism.

At each cycle more error positions are contemporarily evaluated. The risk of such an implementation is in the depth of the paths used to get the following evaluations. An excessive depth brings to an excessive flowing time and/or to an excessive switching activity and therefore to excessive consumption. A slightly different implementation requires the calculation of all the terms directly from the base of the register to limit the depth of the paths.

This implementation increases the number of operators in each RC-logic. All the RC-logic with the same inputs can be grouped in a single RC logic block. In such form, the synthesis tool is able to optimize the net sharing common terms.

Since in the implementation under examination we are operating with a shortened code of N bits, the first $2^{15}-1-N$ positions cannot by definition contain errors and it is therefore useless to evaluate them. To reduce the evaluation time it is therefore possible to initialize the Chien machine with a values $\Lambda_i = \Lambda(\alpha^i L)$, with L given by Eq. (9.51). This can easily be obtained by multiplying all the error locator polynomial coefficients by appropriate constants, α^L, α^{2L}, α^{3L}, α^{4L}, α^{5L}. Such operation can simply be executed by the controller as the last action in the execution of the Berlekamp algorithm. If the size of the message is fixed, the set of constants is univocally determined. If we want a greater flexibility on the dimension of the message, it is essential to count the number of bytes which form the message. For this reason, a counter of the length of the message is inserted in the syndrome computation machine. Such value (N) can be used to identify the set of initialization constants in a set of definite lengths (M).

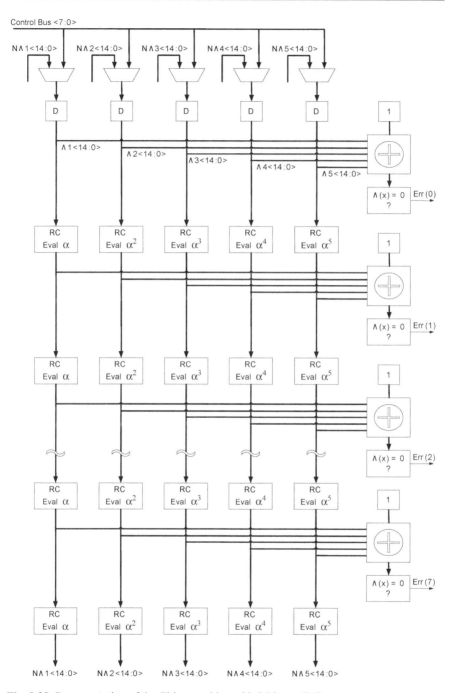

Fig. 9.33. Representation of the Chien machine with 8-bit parallelism

Table 9.2 gives a possible selection.

Table 9.2. Connection between size of the message (N), the range of search of the roots (M) and the value L to be used for the calculation of the respective initialization constants

N	M	L
from 0 to 15 Bytes	16	$2^{15} - 1 - 16$
from 16 to 63 Bytes	64	$2^{15} - 1 - 64$
from 63 to 255 Bytes	256	$2^{15} - 1 - 256$
from 256 to 511 Bytes	512	$2^{15} - 1 - 512$
from 512 to 1023 Bytes	1024	$2^{15} - 1 - 1024$
from 1024 to 2047 Bytes	2048	$2^{15} - 1 - 2048$
from 2048 to 2112 Bytes	2112	$2^{15} - 1 - 2112$

An alternative method to correctly position the Chien machine at the beginning of the message consists in using the value of the counter N to calculate L as:

$$L = 2^{15} - 1 - N \cdot 8 = (l_0, l_1, \ldots, l_{14}) \tag{9.51}$$

Considering that α^L can be expressed as:

$$\alpha^L = \alpha^{2^0 \cdot l_0} \cdot \alpha^{2^1 \cdot l_1} \cdot \alpha^{2^2 \cdot l_2} \cdot \ldots \cdot \alpha^{2^{14} \cdot l_{14}} \tag{9.52}$$

The multiplication by α^L can be broken up into 14 multiplications by definite coefficients. The same can be done to determine α^{2L}, α^{3L}, α^{4L}, α^{5L}. This set of multiplications, executed by the Berlekamp controller, leaves the Chien machine substantially unchanged.

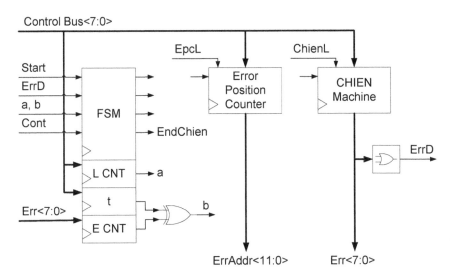

Fig. 9.34. Representation of the Chien machine, first implementation

Figure 9.34 shows a possible implementation of the Chien machine containing: the 8-bit parallelism core described in Fig. 9.33, an "Error Position Counter" to trace the position of the byte under examination, an FSM with a series of support counters to manage the search of error positions. The Error Position Counter is initialized by the controller with a value of N-M, the counter L CNT, which supports the Finite State Machine, is initialized to M. After N-M cycles the Error Position Counter is positioned at zero and the Chien machine has evaluated if there are errors in the first byte. The register t is initialized with the number of errors that the Berlekamp machine estimates to be present in the message and the register E_CNT, initialized to zero, retains the sum of the errors which have been detected. Whenever the Chien machine detects an error, the Error Detect signal (ErrD) is activated to alert the system of the presence of an error. The Chien FSM suspends the search operations leaving the address of the error byte on the bus ErrAddr<11:0> and the mask to be applied for the error correction on the bus Err<7:0>. Once the correction has been performed, the system restarts the FSM through a pulse on the Cont signal. When the Chien machine has examined all the positions or detected all the t errors estimated by the Berlekamp algorithm, it enables EndChien and finishes its function. One or more levels of pipes can simply be inserted into the system to reduce or to avoid the suspension of the Chien machine when the substitution has to be performed.

A slightly more elaborate machine must be used if we want to have addresses and error masks at the end of the process execution by the Chien machine. In Fig. 9.35 there is a schematic representation. An "Error Table Registers" memory, to store the addresses and the error masks, and a register "Error Table Index" under the control of the FSM for the correct addressing of the "Error Table Registers" are added to the Chien machine previously described. The Chien machine so built carries on with the evaluation of all the error positions.

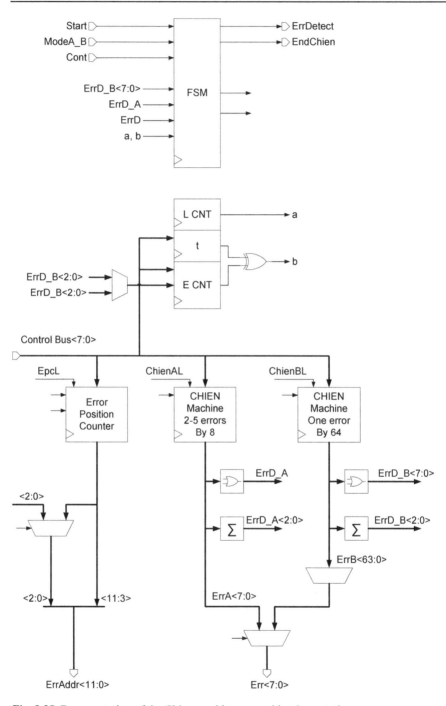

Fig. 9.35. Representation of the Chien machine, second implementation

9.8 Double Chien machine

The execution time of the Chien algorithm is usually seen by the system as an additional time to the information transfer time. If the probability to have one or more errors becomes considerable, the time of search of error positions significantly weights on the system performance. We have seen how a parallel structure of the Chien machine can increase its parallelism and therefore its speed. The increase of the Chien machine parallelism inevitably affects the area used for the BCH machine. The complexity of these machines can be inferred from the number of XOR operators required by the matrices α^1, α^2, α^3, α^4 and α^5 (Fig. 9.32). Being A one of the matrices $\{\alpha^1, \alpha^2, \alpha^3, \alpha^4$ and $\alpha^5\}$ the number of XOR operators needed for this matrix results to be equal to:

$$l = \left(\sum_{i=0}^{14} \sum_{j=0}^{14} a_{i,j} \right) - 15 \qquad (9.53)$$

The matrices $\{\alpha^1, \alpha^2, \alpha^3, \alpha^4$ and $\alpha^5\}$ require $\{1, 2, 3, 4, 5\}$ XOR operators respectively. For the evaluation of an error position we must add the operators for the sum of the Λ_1, Λ_2, Λ_3, Λ_4 and Λ_5 values and the constant "1".

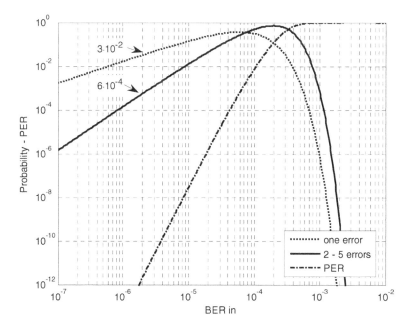

Fig. 9.36. For a 2112-Byte page, representation of single error probability, of 2–5 error probability and of Page Error Rate using an ECC able to correct 5 errors

Altogether, for the evaluation of an error position we need 77 XOR operators. Considering the number of XOR operators present in the various machines for BCH flow, (Chien Machine, SYN Machine, PAR Machine and GF_MUL) we can note that the Chien machine with 8-bit parallelism is responsible for the significant portion of area occupied, exceeding in area the sum of all the other machines.

Figure 9.36 shows, for a 2112-Byte page, the probability of having to correct only one error, the probability of having to correct 2–5 errors and the probability of error (PER) after 5 bits correction as a function of the BER_{in}. For significant values of Page Error Rate (10^{-12}), we have that the probability of a single error is equal to $3 \cdot 10^{-2}$ and the probability of 2–5 errors is equal to $6 \cdot 10^{-4}$ respectively. The probability of a single error is definitely more significant and since the Berlekamp algorithm exactly indicates the number of errors to correct, it may be useful to exploit this information. When we are in the presence of only one error the polynomial $\Lambda(x)$ is a first-degree polynomial.

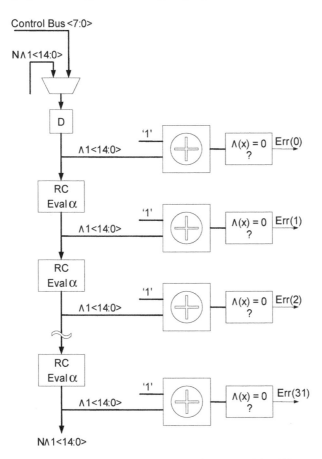

Fig. 9.37. Representation of the simplified core of the Chien machine at parallelism 32 for the search of a single error

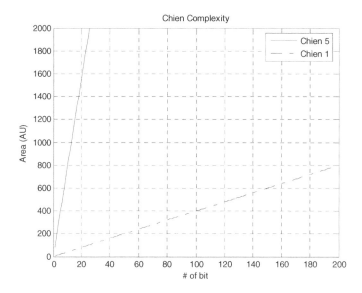

Fig. 9.38. Representation of the complexities for the implementation of the Chien machines 1 (single error) and 5 (2–5 errors)

The machine for the search of the single solution may be reduced to a single element of the original Chien machine (Fig. 9.37) and, at equal complexity, its parallelism can be increased of a factor 8, as indicated by the comparison diagram of the complexities of the two machines (Fig. 9.38).

The system derived is therefore composed of a couple of Chien machines with different parallelisms, one for the correction of the single error and the second for the correction of 2–5 errors (Fig. 9.39).

Such partition, excellent for the system under analysis, has to be re-evaluated and adapted each time. For example, in a hypothetical system which needs to correct 12 errors, the probability of a single error in coincidence with a BER_{in} of interest is too low; it is therefore necessary to distinguish the cases in which 1–2 errors occur from the cases in which 3–12 errors occur, as illustrated in the two diagrams of Figs. 9.40 and 9.41. The comparison of the complexities to realize the Chien machines to correct 1–2 errors vs. 3–12 errors is represented in Fig. 9.42. Also in this case, it turns out to be convenient to divide the Chien machine into more machines with different parallelisms.

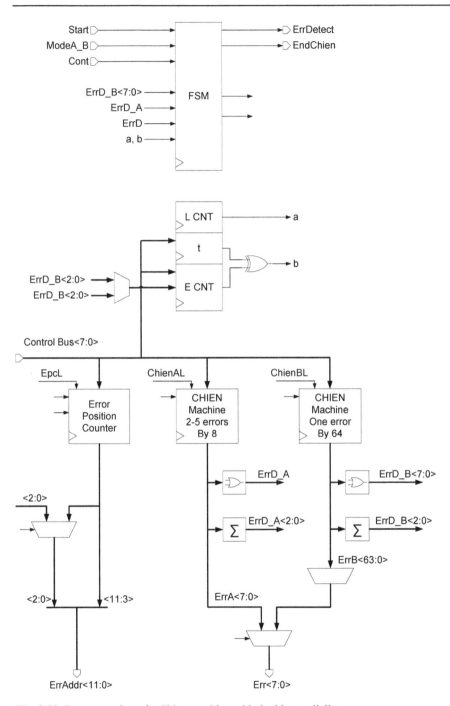

Fig. 9.39. Representation of a Chien machine with double parallelism

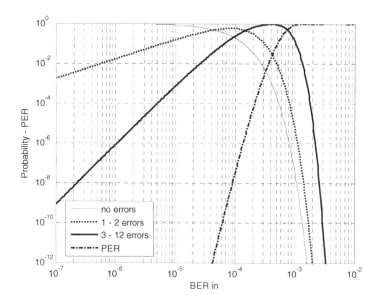

Fig. 9.40. For a 2112-Byte page, representation of the probability of 1–2 errors, of the probability of 3–12 errors and the probability of error on the page (PER) using an ECC able to correct 12 errors

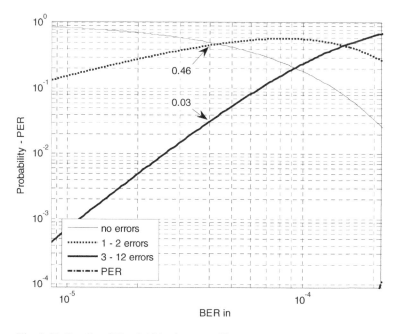

Fig. 9.41. Details of Fig. 9.40 in the area of interest

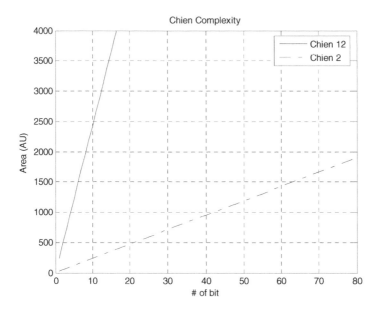

Fig. 9.42. Representation of the complexities for the implementation of the Chien machines 1–2 and 3–12. Further partitions can clearly be considered

9.9 BCH embedded into the NAND memory

Having analyzed how to realize the single parts constituting a BCH machine in detail and with different solutions, let's now see how a selection of these machines can be integrated into a particular practical case. The considered system is a NAND memory with a 2112-byte page and with embedded BCH controller. The logic architecture of a generic NAND memory is functionally represented in Fig. 9.43.

First of all the matrix of a NAND memory is hierarchically organized in blocks constituted by rows which contain two or more information pages. The smallest unit to be programmed and read is the page. The smallest erase unit is the block. The Page Buffer represents the block assigned to the readings and writings of an information page. In the case analyzed the page is constituted by 2112 byte. The Column Decoder provides the addressing of the data contained in the Page Buffer towards the Input Output Buffer (IO) during read operations and, vice versa, in write operations. The Row Decoder selects the block and the matrix row on which to perform the read or modify operations. The Input Output Buffer represents the interface of the NAND towards the outside. The interface signals are substantially: the input and output data bus IO <7:0>; the clock WE# for the input of commands (when CLE is active), of addresses (when ALE is active) and of data (when ALE and CLE are not active); the clock RE# for the reading of data and of the status register; an enabling signal for the device CE# and finally a warning signal for the execution of some algorithm within the device RB#.

Through this interface it is possible to give read, write and erase commands to the device. The Command Interface and Counters block (CI&CNT) recognizes the commands provided by the user, it stores the column, page and block address on which the user intends to operate and control the data input from/towards the Page Buffer. The internal bus Data Bus connects the Page Buffer, the Input/Output Buffer and the Command Interface. The Read, Program and Erase operations are managed by the microcontroller unit which controls the Page Buffer, the matrix and all the high voltage circuits HV required for the execution of the operations (Chap. 4). The microcontroller is not usually involved in data input and data output operations. The Control Bus is substantially a service bus of the microcontroller used to run the device in the execution of read and modify operations. The scheme taken into consideration to integrate the BCH functionalities is the following.

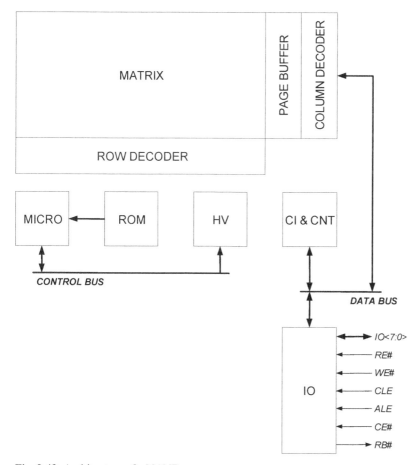

Fig. 9.43. Architecture of a NAND memory

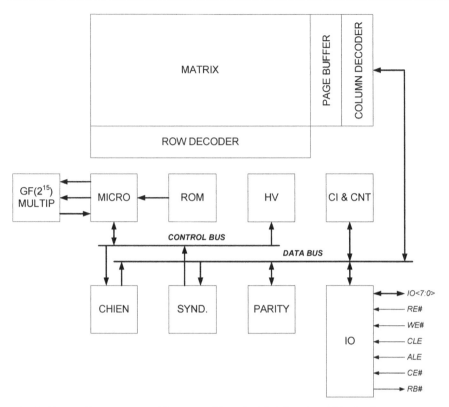

Fig. 9.44. NAND memory architecture with embedded functional blocks for error correction BCH code

The fundamental blocks previously analyzed, for BCH coding and decoding, have been integrated in the architecture (Fig. 9.44). The block for the parity computation (PARITY) is a machine with 8-bit inputs and outputs which is able to receive and elaborate the data that are being loaded towards the matrix. It also has the possibility to pass the values of the parity calculated to be stored after the data on the same page. Since during this phase the only clock available is the clock provided by the user and used for the data input (WE#), the structure of the machine has to elaborate eight bits per cycle. For the decoding we have the syndrome computation block (SYND). Also this one is connected to the Data Bus to be able to calculate the syndromes while the message is taken from the Page Buffer and brought to the output. Besides, the syndrome machine is connected to the Control Bus to be able to pass the syndromes to the microcontroller performing the Berlekamp algorithm. The clock provided by the user during the data output (RE#) is the only active clock during the data output and therefore also the syndrome computation machines have to perform eight steps in a cycle of clock. The bit signaling the presence of one or more errors is sent to the Command Interface to be allocated on a bit of the status register. The user will have to monitor this status bit at the end of the reading of each encoded block. The microcontroller is provided with a multiplier in $GF(2^{15})$, an adder in $GF(2^{15})$, a set of registers and instructions to run the execution of the Berlekamp algorithm. Its working frequency is generally not very elevated and therefore the most suitable GF multiplier is a parallel multiplier. The Chien machine contains the registers required to keep the five error positions and the relative correction masks. This is connected to the microcontroller through the Control Bus, to receive the correct initialization, and it is connected to the Data Bus, to be able to output the processing results. We will now analyze how the system operates and what further changes need to be introduced into the NAND architecture. All the commands sent to the NAND are interpreted by the Command Interface which verifies the syntax and rules the transit of all the information on the Data Bus. Once the Command Interface has recognized a Read, Program or Erase command it activates the microcontroller which executes the proper algorithm. Let's now consider the write command. The typical write command present in NAND memories is represented in Fig. 9.45.

Fig. 9.45. NAND memories standard program command

The command 80h is the initialization command of the program operation, which must be followed by 5 bytes of address, which specify the column address (1 and 2 bytes) and the page address (3–5 bytes) where the data must be programmed. After that, there is a sequence of bytes that is orderly loaded into the Page Buffer and finally the confirmation command 10h after which the CI reawakens the microcontroller, which performs the program operations.

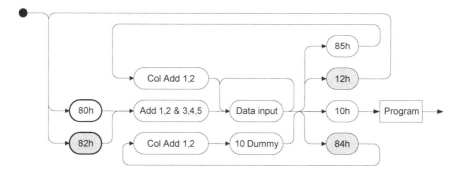

Fig. 9.46. Program command with embedded ECC

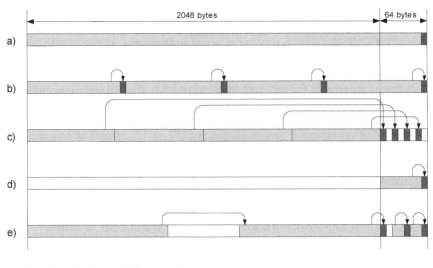

☐ Data Field with ECC protection
■ ECC Parity Bits
☐ Data Field without ECC protection

Fig. 9.47. Some examples of single ECC protected field (**a**), multiple ECC protected fields (**b,c,d**) and mixed ECC protected and not protected fields (**c,d,e**)

During the Data Input phase it is possible to get a random data input command, to change the data insertion point in an absolute way, through the command 85h followed by two column addresses. In this way, the counter, which controls the loading position on the PB, is repositioned in the CI. The CI requires a slightly more articulated command set (Fig. 9.46) to control the parity computation performed by the Parity Machine.

We have assumed, as requisite of the system, that one or more data fields with and without ECC can be inserted on a page. We have also assumed that the data field to be protected is of any length and that we can decide where to position the data part and the parity part on the page with the maximum flexibility. Following the diagram from its point of origin we can simply identify the simple program command, given by the sequence 80h, 5 addresses, 10h. The command 80h is assigned the meaning of program without ECC, the command 82h is assigned the meaning of program with ECC. In both cases we proceed with the addresses and with one data at least. If the data field needs to be protected with ECC (82h) the parity computation block has been activated and it has a parity which must be saved in the matrix. The command 84h activates this mode, the following 10 write cycles, eventually preceded by a column address to specify their locations, transfer the data present in the parity machine to the position specified by the inside counter. Once this parity loading phase is finished, we can proceed to the page program with 10h or to the input of a new data field (12h) without ECC protection (80h) or with ECC protection (82h). With 82h the parity machine is reinitialized ready to generate the parity for the new data field. With this mode it is possible to store a mix of protected and not protected data on the page with the maximum flexibility in positioning the different parts as represented in Fig. 9.47. Let's now consider read operations. In common NAND memories the read command is composed by a sequence represented in Fig. 9.48.

Fig. 9.48. Representation of NAND read command

The command 00h is recognized by the Command Interface as the read command. This must be followed by the column address from which we want to begin to read and by the matrix page address. On the confirm command 30h the Command Interface activates the microcontroller of the device which executes the whole procedure to transfer the information from the matrix to the Page Buffer. The following cycles of data outputs bring the data from the Page Buffer to the output. During this phase it is possible to get an absolute positioning in an absolute way through the command 05h followed by the column address and by E0h. A possible implementation to introduce the ECC functionalities in the read command is represented in Fig. 9.49.

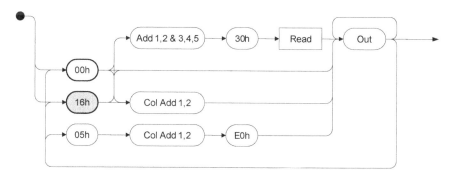

Fig. 9.49. Read command with embedded ECC

The command 16h is added to identify when we want to activate the reading with ECC. This command resets the content of the syndrome computation machines and resets the status flag associated to the syndromes. The read command continues with the part of addresses, the confirmation command 30h and the Read operation, executed by the microcontroller in the usual way. The data of the first field can therefore be drawn through consecutive dataout cycles. If the data field is a field with ECC protection (16h), the syndrome computation machines proceed in parallel adjourning the values at each cycle. The whole data field and the relative parity bits must be taken in order to have valid syndromes values. At the end of the reading of the field with ECC protection, the reading of the status register detects if there is an error in the field. If the error flag is active, it is necessary to use the error position determination command, which is described below, otherwise we can proceed with the reading of the following field: 00h if it is a field without ECC protection, 16h if it is a new field with ECC protection. In both cases a new column address is provided. When, after the reading of a data field with ECC protection, a parity error is detected, we can proceed to the determination of the error positions using the "Get Error Position" (18h) command (Fig. 9.50). The command 18h, recognized by the CI, activates the microcontroller which reads the syndromes, executes the Berlekamp algorithm and determines the most appropriate multiplication coefficients for a fast execution of the Chien algorithm. Then the microcontroller initializes the Chien machine with the values of Λ_1, Λ_2, Λ_3, Λ_4 and Λ_5 and sets all the counters and registers of the Chien machine. Finally, it activates the Chien machine, the complete one if the number of errors is between 2 and 5 and the fast one if only one error has been detected. If the error is not correctable the microcontroller just load "–1" in the register of the number of errors (t) to indicate that the massage contains 6 or more errors and therefore it cannot be corrected. The Chien machine, ruled by a status machine, proceeds to the search of all the error positions saving their locations and their correction masks. At the end of this phase the memory disables the RB# signal to indicate the completion of the operation. Then, the user can proceed with the Get Error Position operation reading the number of errors; 0 if there are no errors to correct, from 1 to 5 if some correctable errors have been recognized, FFh if 6 or more errors have been detected and therefore no correction is possible. The following three read cycles provide the two bytes which identify the position of the first error to be corrected and the relative correction mask respectively. In the same way, the next four 3-Byte sets provide information on other errors. Once the correction of a data field is finished, the data output operation of the following protected and not protected fields can proceed.

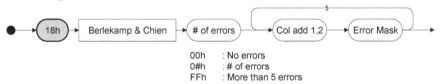

00h   : No errors
0#h   : # of errors
FFh   : More than 5 errors

Fig. 9.50. Representation of the Get Error Position command

In conclusion, in this chapter we have tried to explain how to proceed for the sizing, the designing and the integration of a BCH code in a system based on NAND memories, or similar ones. The possibility to realize machines with elevated parallelism and small complexity, makes the use of BCH codes extremely efficient and effective in the development of storage systems founded on semiconductor memories.

Bibliography

C. Chen, "High Data Rate BCH Encoder" in *US 5040179*, August 1991.

Y. Chen, K. Parthi, "Small Area Parallel Chien Search Architecture for Long BCH Codes" in *IEEE Transactions on Very Large Scale Integration (VLSI) Systems,* Vol. 12, May 2004.

S. T. J. Fenn, M. Benaissa, D. Taylor, "GF(2^m) Multiplication and Division over the Dual Basis", *Transactions on Computers*, Vol. 45, No. 3, March 1996.

R. I. Foster, B. A. Longbrake, "Microprocessor Based BCH Decoder" in *US 4856004*, August 1989.

C. Kraft, "Closed Solution of Berlekamp's Algorithm for Fast Decoding of BCH Codes", *IEEE Transactions on Communications*, Vol. 39, December 1991.

S. Lin, D. J. Costello, "Error Control Coding: Fundamentals and Applications", *F. F. Kuo Editor*, 1983.

Y. Lin et al. "Error Correction Circuit for BCH Codewords" in *US 5416786*, May 1995.

M. Lunelli, R. Micheloni, R. Ravasio, A. Marelli, "Method for Performing Error Corrections of Digital Information Codified as a Symbol Sequence", *US Patent US20050050434*, 2005.

R. Micheloni, R. Ravasio, A. Marelli, "Method and System for Correcting low Latency Errors in Read and Write Non Volatile Memories, Particularly of the Flash Type", *US Patent US20060010363*, 2006.

R. Micheloni, R. Ravasio, A. Marelli, "Method and System for Correcting Errors in Electronic Memory devices", *US Patent US2006005109*, 2006.

R. Micheloni, R. Ravasio, A. Marelli, "Reading Method of a Memory Device With Embedded Error Correcting Code and Memory Device With Embedded Error Correcting Code", *European Patent EPA06425141.6*, 2006.

R. Micheloni et al. "A 4Gb 2b/cell NAND Flash Memory with Embedded 5b BCH ECC for 36MB/s System Read Throughput", ISSCC Dig. Tech. Papers, San Francisco, February 2006.

S. Mizrachi, D. Stopler, "Efficient Method for Fast Decoding of BCH Binary Codes" in *US 2003/0159103*, August 2003.

A. Nozoe et al. "Memory Device and Memory Card" in *US 2002/0054508*, May 2002.

J. Sankaran, D. Hoyle, "Error correction Structures and Method" in *US 2001/0037483*, November 2001.

J. Sankaran, D. Hoyle, "Error correction Structures and Method" in *US 2002/0002693*, January 2002.

T. Tanzawa, T. Tanaka, K. Takeuchi, R. Shirota, S. Aritome, H. Watanabe, G. Hemink, K. Shimizu, S. Sato, Y. Takeuchi, K. Ohuchi, "A Compact On-Chip ECC for Low Cost Flash Memories", *IEEE Journal of Solid-State Circuits*, Vol. 32, No. 5, May 1997.

S. B. Wicker, "Error Control for Digital Communication and Storage", *Prentice Hall*, 1995.

10 Erasure technique

10.1 Error disclosing capability for binary BCH codes

As we will see later in this chapter, the erasure technique is well applied in conjunction with codes owing a high error disclosing capability.

Binary BCH codes are particularly suitable for this approach because, when more than t errors occur, they recognize it without performing erroneous corrections. In any case, an undetectable error pattern, that is an error pattern identical to a codeword, may occur. In this case there are not error correction codes able to detect it and the message is considered as a correct codeword. It is not easy to estimate this undetectable pattern error probability, since it should be necessary to know all the codewords distribution. Anyway, in the following we will not consider this probability but we want to estimate the capability of BCH codes to detect more than t errors, without performing any corrections.

In order to calculate this probability we will follow the decoding procedure described in the previous chapter step by step.

Suppose we have a BCH code in $GF(2^m)$ able to correct t errors. Also suppose that more than t errors occurred.

The first step of the decoding is the syndromes computation. The syndromes will not be linked to the error positions anymore, but they could result as random elements in $GF(2^m)$.

Then, these syndromes will be processed by the Berlekamp tree. There are some branches of the tree followed only if more than t errors occurred. In this case the code is error disclosing. As an example, we analyze the case of a BCH code in $GF(2^{15})$ able to correct five errors (see Chap. 9), giving the following "probability weights":

- 1 if the test is equal to 0;
- $2^{15}-1$ otherwise.

Then, we calculate the disclosing probability P_{ber}, following all the possibility of Fig. 10.1. Every box represents a null test on the variable written inside it. If the tested variable is a null vector, we follow the branch to the left and we give weight 1 to the branch itself; otherwise we move to the right, giving weight $2^{15}-1$ to the next branch. All the formulas needed and the error locator polynomials are reported in Eq. (9.47) and Eq. (9.48). The disclosing branches are indicated in the tree as ERR boxes: this means that those branches are followed only if more than 5 errors occurred.

All the probabilities for possible disclosing branches that took us to an ERR box have to be added. For example, the first ERR block is reached if S_1, S_3, S_5 and S_7 are equal to 0 while S_9 is different from 0, that is:

$$p1 = 1 \cdot 1 \cdot 1 \cdot 1 \cdot \left(2^m - 1\right) = 2^m - 1 \tag{10.1}$$

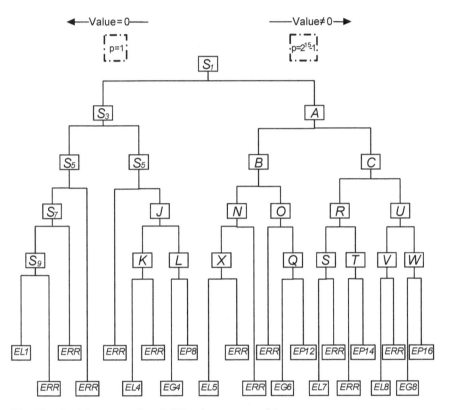

Fig. 10.1. Berlekamp tree for a BCH code corrector of 5 errors

The second ERR block is reached if S_1, S_3, S_5 are equal to 0, S_7 is different from 0, while there are no constraints on S_9, that is:

$$p2 = 1 \cdot 1 \cdot 1 \cdot \left(1 + 2^m - 1\right) \cdot \left(2^m - 1\right) = 2^m \left(2^m - 1\right). \tag{10.2}$$

The lack of constraints is represented by the sum $1 + 2^m - 1$, which represents the total number of the field's elements. In this way all the probability contributors have been found out.

The sum of all these probabilities must be divided by the number of the field's elements, taken 5 times. It can be shown that the overall probability is asymptotically equal to P_{ber}

$$P_{ber} = \frac{1}{2^m} \qquad (10.3)$$

This probability is very small, so it is possible to state that, when more the t errors occur, the Berlekamp tree algorithm is usually unable to disclose it and a wrong branch is followed.

Using the approach described above, it is possible to find the probability of following the 0, 1, 2, 3, 4 and 5 errors branches. Naming P_0 the probability of following the 0 errors branch, P_1 the probability of following the single error branch and so on, we calculate the terms:

$$P_0 = \frac{1}{\left(2^m\right)^5} \qquad (10.4)$$

$$P_1 = \frac{2^m - 1}{\left(2^m\right)^5} \qquad (10.5)$$

$$P_2 = \frac{\left(2^m - 1\right)^2}{\left(2^m\right)^5} \qquad (10.6)$$

$$P_3 = \frac{\left(2^m - 1\right)^2 + \left(2^m - 1\right)^3}{\left(2^m\right)^5} \qquad (10.7)$$

$$P_4 = \frac{2 \cdot \left(2^m - 1\right)^3 + \left(2^m - 1\right)^4}{\left(2^m\right)^5} \qquad (10.8)$$

$$P_5 = \frac{3 \cdot \left(2^m - 1\right)^4 + \left(2^m - 1\right)^5}{\left(2^m\right)^5} \qquad (10.9)$$

As it is possible to see, P_5 is asymptotically equal to 1, which means that when more then 5 errors occurred, quite most of the times, the branch taken by the Berlekamp tree is the one concerning five errors.

In this case the Chien machine is the step of the decoding algorithm that has to disclose the errors. Now we have a polynomial which seems right and we find the probability that it has separate roots in the field and that they are in the first N

positions, where N is the codeword length. In this case the ECC is not able to disclose the errors and it makes erroneous corrections, introducing new errors.

Supposing we have a polynomial of degree t, i.e. a five errors branch of the Berlekamp tree has been reached, the probability P_{pol} of finding t separate roots in the field is given by:

$$P_{pol} = \frac{\binom{2^m}{t}}{\left(2^m\right)^t} \tag{10.10}$$

The BCH codes usually used in applications are the shortened BCH codes. In this case we also have to consider the probability of finding the roots in the positions corresponding to a valid location in the code:

$$P_{shor} = \left(\frac{N}{2^m}\right)^t \tag{10.11}$$

Finally, the error probability of t-error correcting and more than t-error disclosing BCH code is given by

$$P = \left(1 - \sum_{i=0}^{t}\binom{N}{i}(1-p)^{N-i}\,p^i\right) \cdot \left(1 - \frac{\binom{2^m}{t}}{\left(2^m\right)^t}\left(\frac{N}{2^m}\right)^t + \frac{1}{2^m}\right) \tag{10.12}$$

On the other hand, the error probability of having BCH t-error correction code unable to disclose more than t errors is given by:

$$P = \left(1 - \sum_{i=0}^{t}\binom{N}{i}(1-p)^{N-i}\,p^i\right) \cdot \left(\frac{\binom{2^m}{t}}{\left(2^m\right)^t}\left(\frac{N}{2^m}\right)^t - \frac{1}{2^m}\right) \tag{10.13}$$

The chart obtained plotting the probability of Eq. (10.12), Eq. (10.13) and the error probability of a five-error BCH correction code, is reported in Fig. 10.2.

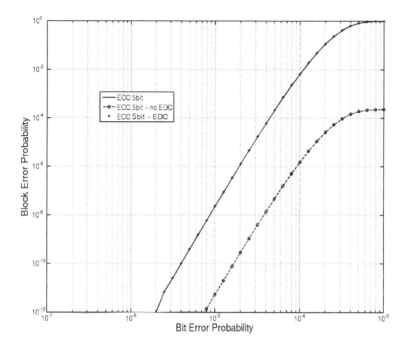

Fig. 10.2. Comparison between BCH codes with or without disclosing capability

As it possible to see, the curves of the 5-error ECC and 5-error ECC code with EDC capability overlap, meaning that a BCH code is also an error disclosing code in most of the cases. On the other hand, the probability of having an error correcting code without disclosing capability is very low.

In this way it is possible to see that BCH codes usually have very good disclosing capability, that is, when more than t errors occurred, the code does not perform erroneous corrections.

10.2 Erasure concept in memories

In digital communication systems, the information is formatted into sequences of symbols which can take a value only from a finite set of possibilities. In the case of a binary system, this set is formed only by 0s and 1s. The receiver examines the message received and decides, among all the possible symbols transmitted, those which have most probably been sent. As seen in Sect. 1.3, the decoding is based on a threshold decision circuit.

In a hard-decision receiver, the detection circuit limits its range of choices to the same group of choices available to the transmitter. In the case of a binary system, also the receiver's possibilities of choices are between the symbols 0 and 1. In some cases, however, the signal received does not give a clear indication of which symbol has most probably been transmitted. Rather than forcing the decision, a soft-decision receiver uses a selection of choices to communicate the quality of the received symbol to the decoder. The simplest form of soft-decision receiver uses erasures to indicate that it is in doubt about a certain received symbol.

Accordingly, as far as the binary channel is concerned, while the transmitter can only choose between the values 0 and 1, the receiver can choose among the values 0, 1 and X, which is used to indicate that the symbol was not understood. Erasure decoding does not provide significant additional gain for additive white Gaussian noise channels, but it does provide substantial improvement over fading and bursty channels.

So far, through the application of a correction code to memories, we have understood how the memory failure probability can decrease. However, every type of error correction codes described in the previous chapters concerns the correction of hard fails. The erasure concept, and the way it is described in telecommunications, is not applicable to memories, since it never happens not to understand a symbol. With reference to Fig. 10.3, the hard decision-maker, which uses a discriminating threshold (in the figure the threshold is 0), "reads" the threshold voltage of a cell and compares it with its own threshold in order to establish if the read voltage is greater or less than 0: if it is less then 0, the decision maker decodes a 1, otherwise a 0. For more details regarding the read operation in NOR and NAND memories, refer to Chap. 3 and Chap. 4.

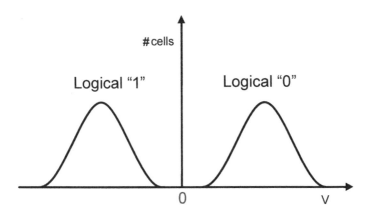

Fig. 10.3. Voltage distributions per memory cells

Actually, as already explained in Chap. 5, the nonvolatile memory devices are subject to deterioration with use, so that the number of errors grows with time. In Flash memories, both NAND-type and NOR-type, the cell oxide degrades both for time and for electric solicitations. Accordingly, the threshold voltage distributions associated to the program levels get closer and overlap. Besides, the threshold decision-maker itself is not able to discriminate the threshold voltage of a cell in a digital way, but, as for all the instruments, it is subject to uncertainty errors.

In conclusion, as we can see in Fig. 10.4 the cells distributions have considerably got closer as far as to overlap. Besides, there is a grey Gaussian which represents the instrument uncertainty, that is, the distribution of the possible errors due to the measuring instrument.

The following cases may occur.

- The cell in area *A*. This area is distant from the Gaussian of the instrument uncertainty and therefore the cell is read as 1.
- The cell in area *B*. This area is distant from the Gaussian of the instrument uncertainty and therefore the cell is read as 0.
- The cell in area *C*. This area is distant from the Gaussian of the instrument uncertainty and therefore the cell is read as 0. The read value is, however, erroneous and the cell is incorrectly read, because the distributions have overlapped.
- The cell in area *D*. This area is distant from the Gaussian of the instrument uncertainty and therefore the cell is read as 1. The read value is, however, erroneous and the cell is incorrectly read, because the distributions have overlapped.
- The cell in area *E*. This area coincides with the area of uncertainty of the instrument and therefore the read value can randomly be either a 0 or a 1.

Obviously, the Gaussian of the instrument uncertainty should occupy a very small area. It happens, however, that this area may increase in dimensions as much as to become significant. In fact, as the distributions shift of the threshold voltages appears, the sensibility to random fluctuations grows and, accordingly, also the probability that two consecutive readings of the same cell, in the same configuration, will give two different results grows.

There is a source of errors known as *Random Telegraph Noise* (RTN), which has as an effect the impossibility to precisely establish the charge state of the cell, so that two readings under the same conditions could output different results.

Using this characteristic of Flash memories, we can introduce the erasure concept. It has been said that an erasure is detected whenever consecutive readings of the same cell under the same conditions give different results.

Substantially, this erasure concept applied to memories is equivalent to modifying the threshold decision-maker by replacing a single threshold with two uncertainty thresholds.

Obviously also these two thresholds suffer from uncertainty errors, due to the fact that the read operation is always based on a threshold decision-maker. Anyway, if the area between $[-S, +S]$ is big enough, these errors are negligible compared to RTN errors. As we see in Fig. 10.5, there may be three possible scenarios:

- the voltage of the read cell is inferior to $-S$, then the cell is read as a 1;
- the voltage of the read cell is superior to $+S$, then the cell is read as a 0;
- if the voltage of the read cell is included between $-S$ and $+S$ then the cell is read as an erasure.

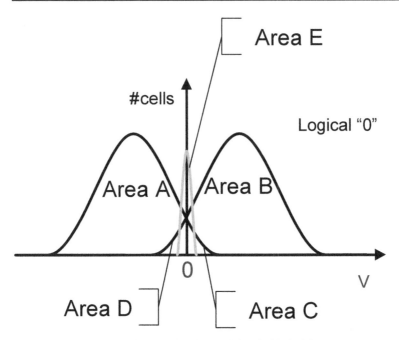

Fig. 10.4. Distributions after deterioration and threshold decision

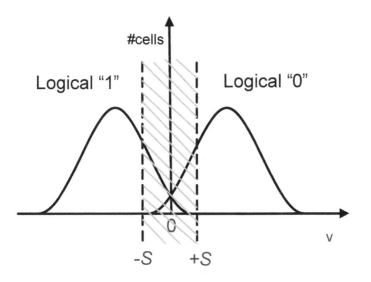

Fig. 10.5. Erasure applied to the threshold decision-maker

Observe that throughout this chapter, the uncertainty interval is presented as symmetrical, even if this could not be always true. In particular, if the error source is asymmetrical we can choose two different thresholds in order to better protect ourselves from the most probable error.

Later in this chapter we will see how the use of a decoding with erasure can lead to an extension of the device life, but, first, it is necessary to understand what we mean by decoding with erasure.

10.3 Binary erasure decoding

In Chap. 1 we introduced the basic code parameter d_{min}, representing the minimum distance. Suppose we have messages with a single erasure. The codewords have minimum distance $d_{min}-1$ among the un-erased coordinates. In general, if f erasures occur, the codewords will have minimum distance $d_{min}-f$ among the un-erased coordinates. Accordingly, the code is able to correct

$$t_e = \left\lfloor \frac{d_{\min} - f - 1}{2} \right\rfloor \qquad (10.14)$$

errors between the un-erased coordinates. The square brackets represent the floor function.

In other words, it is possible to correct e errors and f erasures satisfying Eq. (10.15).

$$2e + f < d_{\min} \qquad (10.15)$$

Observe that we can correct a number of erasures twice bigger than the number of errors. The reason is that, for erasures, the position is already known.

Now, we will deal only with binary decoding with erasures. This can be performed with very few changes in comparison with a standard decoding. The procedure is described here below.

- Once the word r has been received, substitute all erasures with 0s and decode with the standard algorithm. The resultant codeword is c_0.
- Substitute all erasures with 1s and decode with standard procedures. The resultant codeword is c_1.
- Compare c_0 and c_1 with r and select the codeword that has the minimum Hamming distance from r.

Observe that it could be erroneous to deduce that the substitution of erasures with 0s or 1s is performed randomly. In reality, the substitution, first with all 0s and then with all 1s, ensures the correct running of decoding operations.

First, suppose that the number of errors and erasures occurred satisfy Eq. (10.15). By substituting all erasures with 0s we can generate e_0 errors, so that there could be $e+e_0$ errors at the input of the decoder. On the other side, by substituting all erasures with 1s we could have $e+e_1=e+f-e_0$ errors at the input of the decoder. Either e_0 or $f-e_0$ must be at most equal to $f/2$. It follows that, at least for one decoding operation, the total number of errors e_t satisfies

$$2e_t \leq 2\left(e+\frac{f}{2}\right) < d_{\min}$$ (10.16)

Accordingly, at least one decoding operation takes to the correct codeword.

In the case of memories, we first identify the erasures. The erasure decoding process is useful when used in cascade with an error correction code embedded in a memory. The purpose is to try to correct errors also when the embedded code correction capability has been overcome. For this reason, at first we try to perform a classic error correction; in case of fail we proceed with the erasure process. It is however necessary to have a code with quite high disclosing capability, since the code must be able to signal when its error correction capability has been overcome, without performing erroneous corrections. Cyclic codes are very good for this purpose as described in Sect. 10.1.

The procedure plans a first read of the message and a decoding with the embedded error correction code. If the classic decoding records a failure, we proceed with a second reading under the same conditions of the previous one.

If we are in one of those cases where the result of two different readings takes to random values, there will be cells which at the first reading will have been read as 1s and at the second as 0s or vice versa. All those cells are marked as erasures. At this point the binary erasure algorithm described above can be performed. As shown in Fig. 10.6, the key of the process consists in performing a comparison bit by bit between the first and the second read values. In the case of two readings the comparison bit by bit can be performed through an exclusive-OR operator.

Once the comparison bit by bit is performed, the string containing the symbols 0, 1 and X is stored and can be decoded.

In conclusion, the re-reading process makes advantageous use of the detection of cells which have incoherent results during reading. These positions are marked as erasures, thus increasing the probability of the decoding to be successful, after the failure of the first decoding with a conventional error correction code.

It is also possible to perform a greater number of readings, thus increasing the probability of recognizing fluctuations. There will be, in fact, cells for which the shift of the threshold voltage has brought them to the reference point: hence, they will be read half of the times as a 1 and half of the times as a 0, but there will also be cells for which the shift has not been so dramatic, hence, the cell is read 20% of the times as a 1 and 80% of the times as a 0.

Increasing the number of the readings also helps to identify the cells, whose threshold voltage distribution is not exactly halfway through the reference point. With reference to Fig. 10.5, increasing the number of readings corresponds to increasing the threshold value $+S$ (and to decreasing the threshold value $-S$), thus widening the uncertainty interval.

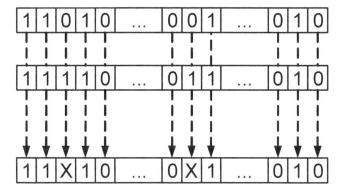

Fig. 10.6. Erasures detection for two consecutive readings

In the case of more readings, it is necessary to find a way to compare all the readings bit by bit. With reference to Fig. 10.7, we perform an arithmetic sum, bit by bit, of all the readings and we store the resulting string into an auxiliary memory register. Then, this string is analyzed in order to construct the string S_M needed by the decoding algorithm.

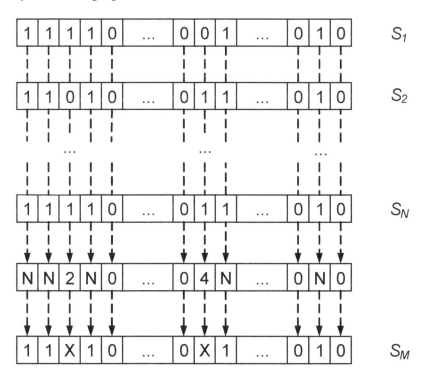

Fig. 10.7. Erasures detection for a number N of re-readings

In practice, once N readings are performed, one of the following cases can occur on each bit:

- The sum of all the readings gives 0 as a result; in this case all the readings have given 0 as a result, thus the string S_M contains a 0.
- The sum of all the readings gives N as a result; in this case all the readings have given 1 as a result, thus the string S_M contains a 1.
- The sum of all the readings gives a number between 0 and N as a result; in this case not all the readings have been coherent and the cell has sometimes been read as a 0 and sometimes as a 1. In this case the string S_M will contain an erasure, marked with an X.

The erasure process is useful to take care of all random error effects. As already said, RTN can benefit from the erasure process. The RTN causes a shift of the cell threshold voltage of several hundreds milliVolt. That value, since it depends on the technology and cannot be removed, could lead to errors especially in the case of multilevel memories where the distributions are closer. Considering the random behavior of this kind of error, an erasure decoding is well suited for the purpose.

10.4 Erasure and majority decoding

As explained in the previous section, the probability to detect an erasure raises, increasing the number of readings, since it is more likely to see the least frequent phenomena.

On the other hand, however, increasing the number of readings may lead to the detection of a number of erasures such that Eq. (10.15) cannot be satisfied anymore.

It is thus possible to combine the erasure process with the majority process, in order to mark with erasures only the cells for which several transitions have been detected and to decode by majority the cells that have shown less frequent transitions.

For this reason, we introduce two thresholds s_1 and s_2. Suppose we perform N readings of the message m, the vector S_M at the input of the decoder will be constructed as follows:

- If, for a given cell, the sum of all the N readings gives a number a at most equal to s_1 as a result, then the corresponding cell in S_M will contain a 0;
- If, for a given cell, the sum of all the N readings gives a number b grater than or equal to s_2 as a result, then the corresponding cell in S_M will contain a 1;
- in all the other cases the corresponding cell in S_M will contain an erasure.

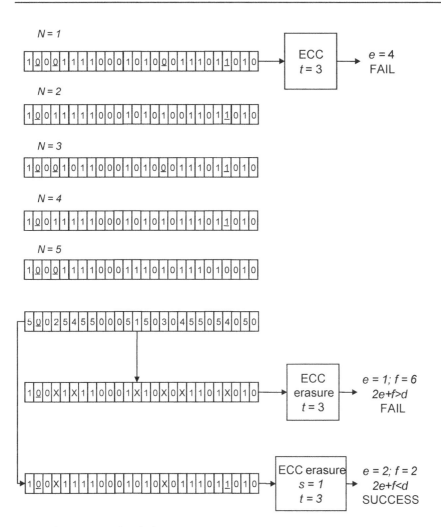

Fig. 10.8. Erasure and majority process

The description will proceed taking into account symmetrical intervals with only one threshold Ns, so that $s_1=Ns$ and $s_2=N–Ns$.

With reference to Fig. 10.8, note how the combined erasure and majority decoding is successful compared to the simple decoding with erasure.

In this case a 3-error correction code is considered, whose minimum distance is 7, and we perform 5 readings. The figure shows the first reading, which gives failure, since there are 4 errors in the underlined bits. The 5 consecutive readings are indicated, as well as the vector sum.

Using an erasure decoding procedure we still have an error and, besides, we detect 6 erasures, so that Eq. (10.15) is not valid anymore and the decoding fails.

Instead, an erasure decoding with a threshold for the majority decision introduces one more error but it eliminates 4 erasures so that Eq. (10.15) is satisfied and the decoding is successfully completed.

As written at the beginning of this chapter, the erasure decoding process is useful when used together with a code embedded in the memory.

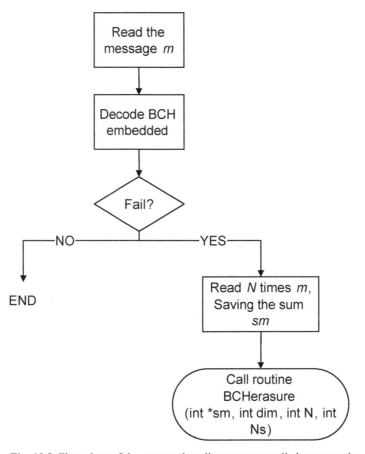

Fig. 10.9. Flow chart of the erasure decoding process applied to memories

The example described above follows the general flow chart given in Fig. 10.9, where *sm* indicates the memory space where the sum vector is allocated, *N* indicates the number of readings performed, *Ns* indicates the threshold in the case of majority decoding (it can be null if we want to use a decoding with only erasure) and *dim* is the number of bytes of the message.

The erasure strategy is particularly useful when the ECC embedded is a BCH code because of their great error disclosing capability (Sect. 10.1).

In this case, the software routine follows the flow chart of Fig. 10.10 leading the decoding to success or failure.

In Appendix E a possible implementation of the BCH erasure routine is described.

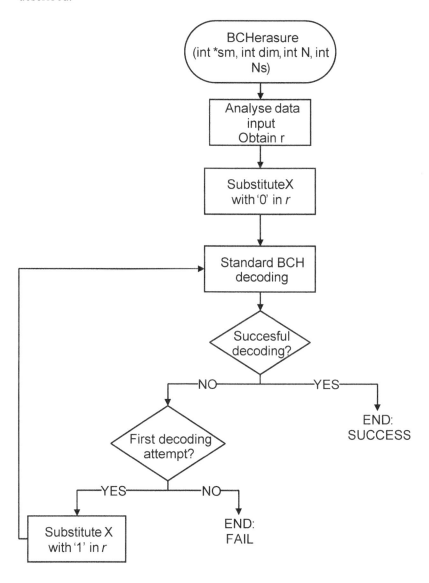

Fig. 10.10. Flow chart of the erasure decoding process for a binary BCH code

10.5 Erasure decoding performances

The erasure decoding process results to be very useful when time and use make the shift of the threshold voltages unavoidable. In that case, consecutive readings of the same cell under the same conditions may lead to different results. The success of the erasure decoding is based on the ability of discovering incoherent results.

Estimating the failure probabilities is not simple, because we need to assign a probability to the hard error and a probability to the erasure detection. The chart of Fig. 10.11 shows the failure curves for a standard 5-error correction BCH code, for a BCH code with erasure corrector of 5 errors and for a BCH code with erasure and majority threshold fixed at 1. Number of readings is fixed.

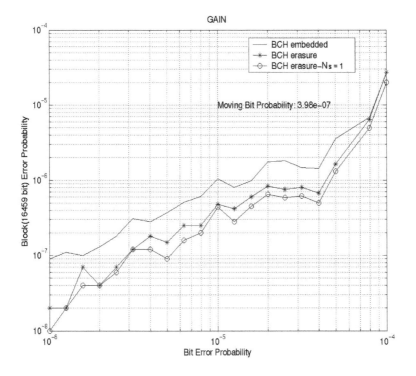

Fig. 10.11. Failure curves for a standard BCH code, a BCH code with erasure and a BCH code erasure + majority

The chart fixes the probability of detecting an incoherency in the readings at a certain value and the results are obtained through simulations.

As we can note, in general, the erasure process is better than the standard BCH decoding process. When error probabilities are low enough, the erasure and the majority+erasure processes can be considered equivalent. As the error probability grows the majority+erasure process results to be better, since it introduces a smaller number of erasures that could invalidate Eq. (10.15).

The chart in Fig. 10.12 represents the effect of increasing the number of readings in the erasure process. Generally, the increase is a benefit, because more "moving cells" can be detected. Anyway, if the bit error probability is too high this increase leads to a decoding failure because Eq. (10.15) is not satisfied anymore. As already said, in this case the introduction of a threshold decreases the error probability.

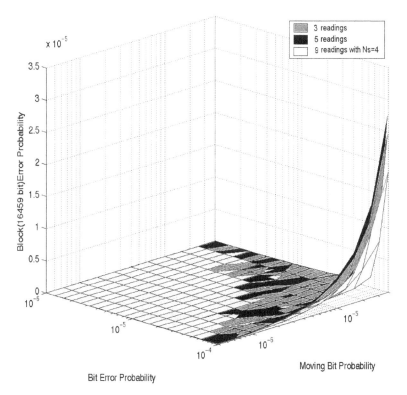

Fig. 10.12. Failure planes for a BCH erasure code using respectively 3 read, 5 read and 9 read with the threshold $s=4$

The chart is a 3D chart because there are 3 error probabilities to take into account:

- the hard fail bit error probability;
- the moving bit error probability;
- the chip error probability.

In all these cases the error correction code considered is a BCH code able to correct 5 errors.

If, on one side, the erasure decoding process considerably improves the error probability, on the other side there is an increase in the decoding time and in the read cycle time needed to detect the erasures, which may became N times longer. In fact, once the erasures have been detected, it is necessary to perform the decoding, which, in the best case, is successful at the first attempt, that is, by substituting the erasures with 0s, while, in the worst one it is necessary to perform the decoding by substituting the erasures with 1s. Therefore, the decoding time may be doubled or tripled.

Instead, for computational complexity, the decoding process with binary erasure does not introduce new concepts, since it can be obtained with small changes in the standard decoding algorithm. Thus, the computational complexity does not increase.

In general, the erasure process just described is applied to memories in order to extend the device life when the embedded error correction code is not able to guarantee reliable information anymore.

Bibliography

E. R. Berlekamp, "The Construction of Fast, High-Rate, Soft Decision Block Decoders" in *IEEE Transactions on Information Theory*, Vol. 29, May 1983.

G. Einarsson, C. E. Sundberg, "A Note on Soft Decision with Successive Erasures" in *IEEE Transactions on Information Theory*, January 1976.

P. Fitzpatrick, "New Time Domain Errors and Erasures Decoding Algorithm for BCH Codes" in *Electronics Letters*, Vol. 30, January 1994.

P. Fitzpatrick, "Errors-and-Erasures Decoding of BCH Codes" in *IEEE Proceedings on Communications*, Vol. 146, April 1999.

V. Intini, A. Marelli, R. Ravasio, R. Micheloni, "Reading Method of a Memory Device with Embedded Error-Correcting Code and Memory Device With Embedded Error-Correcting Code" in *US 2007/ 0234164*, October 2007.

N. Kamiya, "On Acceptance Criterion for Efficient Successive Errors-and-Erasures Decoding of Reed-Solomon and BCH Codes" in *IEEE Transactions on Information Theory*, Vol. 43, September 1993.

N. Kamiya, "On Algebraic Soft-Decision Decoding Algorithms for BCH Codes" in *IEEE Transactions on Information Theory*, Vol. 47, January 2001.

O. Keren, S. Litsyn, "A Class of Array Codes Correcting Multiple Column Erasures" in *IEEE Transactions on Information Theory*, Vol. 43, November 1997.

M. G. Kim, J. H. Lee, "Undected Error Probabilites of Binary Primitive BCH Codes for Both Error Correction and Detection" in *IEEE Transactions on Communications*, Vol. 44, May 1996.

R. Koetter, A. Vardy, "Algebraic Soft-Decision Decoding of Reed-Solomon Codes" in *IEEE Transactions on Information Theory*, Vol. 49, November 2003.

H. Kurata, K. Otsuga, A. Kotabe, S. Kajiyama, T. Osabe, Y. Sasago, S. Narumi, K. Tokami, S. Kamohara, O. Tsuchiya, "The Impact of Random Telegraph signals on the Scaling of Multilevel Flash Memories" in *IEEE Symposium on VLSI Circuits Digest of Technical Papers*, 2006.

H. Tanaka, K. Kakigahara, "Simplified Correlation Decoding by Selecting Possible Codewords Using Erasure Information" in *IEEE Transactions on Information Theory*, Vol. 29, September 1983.

H. Tanaka, K. Kakigahara, "Simplified Correlation Decoding by Selecting Possible Codewords Using Erasure Information" in *IEEE Transactions on Information Theory*, Vol. 29, September 1983.

T. L. Tapp, A. A. Luna, X. Wang, S. B. Wicker, "Extended Hamming and BCH Soft Decision Decoders for Mobile Data Applications" in *IEEE Transactions on Communications*, Vol. 47, March 1999.

A. Vardy, Y. Be'ery, "Maximum-Likelihood Soft Decision Decoding of BCH Codes" in *IEEE Transactions on Information Theory*, Vol. 40, March 1994.

S. B. Wicker, "Error Control for Digital Communication and Storage", *Prentice Hall*, 1995.

Appendix A: Hamming code

This Appendix is a collections of routines written in C++ language that find the different parity matrices described in Chap. 7.

Every routine is composed of different subroutines that will not be described in details.

A.1 Routine to find a parity matrix for a single bit or a single cell correction

In the following a C++ routine finds possible parity matrices suitable for the correction of a single bit or a single cell. The parity matrix is built following the rules of Chap. 7, Sect. 7.2.1. All the results are written on a file. A particular attention must be paid to the number of check bits in case of a cell error correction. This number must be even in order to avoid a partially empty cell. If this number is odd an all 0s row must be added.

INPUTS

- the file where the results must be written;
- the number of data bits;
- the number of check bits;
- the maximum number of 1s per row;
- if we want to correct a bit or a cell;

OUTPUTS

- it rewrites the number of data bits
- it writes the minimum number of check bits used to find a code;
- the maximum number of 1s per row for the found code;
- the code and, if we want to correct a cell, the sum of 2 adjacent rows that must be skipped according to the imposed constraints.

ROUTINE

```
#include <iostream.h>
#include <stdio.h>
#include <conio.h>

/* Service routines */
int nbitf(int n);
int add(int a, int b);
void stampa(char buf[], int x, int m);
void scrivo(char buf[]);
void aggiu(char a[], char b[]);
int len(char b[]);
void ouconv(char a[], int b, int c);
int vabs(int a);
int norma(int n, int a[], int c[]);
void matr(int n, int a[], int b[10][10]);
int pause();

FILE *sto;

int main() {
    int i, j, k, t = 0, nbitp, npa, n512, nbit, 164,inizio;
    char nomeo[80], bu[80];
    int m64[2049], gen[2049];
    int ngen, m121[2048], kr, ks, pa, nfine, celle;

    cout << "file for the results "; cin >> nomeo;
    /* it asks where it has to write the results */
    Sto = fopen(nomeo, "wb"); if(sto == 0) return(0);

1105:
    cout << "codeword length [0 =end] "; cin >> nbit;
    /* it asks how many data bits it has to consider */
    if(nbit == 0) {fclose(sto); return(0);}
    cout << "number of check bits "; cin >> nbitp; n512=1 << nbitp;
    /* it asks how many check bits it has to search for */
    cout << "maximum number of 1s per row "; cin >> npa;
    /* it asks for the maximum number of 1s per row */
    cout << "cell correction [1=yes 0=no] "; cin >> celle; 164 = 0;
    /* it asks if we want to correct one bit or one cell */
    for(k = 1; k <= npa; k++){
        /* it generates all the possible words */
        for(i = 0; i < n512; i++){
            if(nbitf(i) == k){
                m64[164] = i; 164++;
            }
        }
    }
    Ngen = 0; ks = 0; kr = 0;
    Nfine = nbitp + nbit; inizio = nbitp;
    /* this is the matrix to be found */
    /* Identity matrix */
    for(i = 0; i < nbitp; i++) {
        pa = m64[i];
        if(((kr % 2) == 1) && (celle == 1)) {
        /* if it has to correct the cell and the position is odd it
sums the previous row */
```

```
      /* it puts on a forbidden-words matrix the sum */
         gen[ngen] = add(pa, m121[kr - 1]); ngen++;
      }
      /* it chooses the word and marks the forbidden-word table */
      m121[kr] = pa; kr++; gen[ngen] = pa; ngen++;
   }
   if(((nbitp & 1) == 1) && (celle == 1)) {
      /* we must have an even number of check bits, so if the num-
ber is odd we put an all 0 row */
      nfine++; inizio++; m121[kr] = 0; kr++;
   }

l100:
   ks = nbitp; /* it begins the search of the matrix */
l11:
   pa = m64[ks];
   for(j = 0; j < ngen; j++) {
      if(gen[j] == pa) { /* the word is forbidden */
         ks++; /* it chooses another word */
         if(ks < 164) goto l11; /* if another word exists in the
possible choices */
         else goto l200; /* It's finished and it hasn't found a
code */
      }
   }
   if(((kr % 2) == 1) && (celle == 1)) {
      /* it has to correct the cell and the position is odd */
      /* it looks if the sum with the previous row is a forbidden
word */
      t = add(pa, m121[kr - 1]);
      for(j = 0; j < ngen; j++) {
         if(t == gen[j]) {
            ks++;
            if(ks < 164) goto l11;
            else goto l200; /* it has to choose another word, if
it is possible */
         }
      }
      /* it puts the sum in the forbidden-words table */
      gen[ngen] = t; ngen++;
   }
   /* it chooses the word and marks the forbidden-word table */
   m121[kr] = pa; kr++; gen[ngen] = pa; ngen++;
   if(kr == nfine) { /* it has finished */
      bu[0] = 0; aggiu(bu, "data results ="); ouconv(bu, nbit, 5);
      aggiu(bu, " parity bits ="); ou - conv(bu, nbitp, 5);
      aggiu(bu, " uni =");
      ouconv(bu, npa, 5); scrivo(bu);
      for(i = inizio; i < kr; i++) {
         bu[0] = 0; ouconv(bu, i + 1 - inizio, 4);
         aggiu(bu, "  "); stam - pa(bu, m121[i], nbitp);
      } /* it writes the matrix */
      if(celle == 1) {
         bu[0] = 0; aggiu(bu, "cells matrix"); scrivo(bu);
         for(i = inizio; i < kr; i = i + 2) {
            bu[0] = 0; ouconv(bu,(i - inizio) / 2 + 1, 4);
            aggiu(bu, "  ");
            t = add(m121[i], m121[i + 1]); stampa(bu, t, nbitp);
```

```
            }
 /* it writes the sum of two rows (0-1, 2-3, ...) if it has to cor-
rect the cell */
        }
        goto 1105;
    }
    if(ks < 164) goto 1100;
1200:
    bu[0] = 0; aggiu(bu, "matrix not found ="); ouconv(bu, nbit, 5);
    aggiu(bu, " parity ="); ouconv(bu, nbitp, 5);
    aggiu(bu," ones ="); ouconv(bu, npa, 4);
    scrivo(bu); goto 1105;
}

int pause() {
    int k;
    k = getch();
    return(k);
}

void matr(int n,int a[],int b[10][10]) {
    int i, j, t, k;
    for(i = 0; i < 8; i++) {
        for(j = 0; j < 8; j++) {
            t = 0;
            for(k = 0; k < n; k++) {
                if(((a[k] & (1 << i)) > 0) &&
                   ((a[k] & (1 << j)) > 0))
                    t++;
            }
            b[i][j] = t;
        }
    }
}

int norma(int n, int a[], int c[])  {
    int r = 0, i, t, k, rm = 1000;

    for(i = 0; i < 8; i++) c[i]=0;
    for(i = 0; i < n; i++) {
        t = a[i]; k=0;
11:     if((t % 2) == 1) c[k]++;
        k++; t = int(t / 2); if(t > 0) goto 11;
    }
    for(i = 0; i < 8; i++) {
        if(r < c[i]) r = c[i];
        if(rm > c[i]) rm = c[i];
    }
    r = r * 1000 - rm;
    return(r);
}

void scrivo(char bu[]) {
    int l;
    l = len(bu);
    cout << bu << endl;
    bu[l] = 13; bu[l + 1] = 10; bu[l + 2] = 0;
    fwrite(bu, l + 2, sizeof(char), sto);
```

```
}

int vabs(int a) {
    int y;
    y = a;
    if(y < 0) y = -y;
    return(y);
}

int nbitf(int n) {
    int y = 0;

l1: if((n & 1) > 0) y++;
    n = int(n / 2); if(n > 0) goto l1;
    return(y);
}

int add(int a,int b) {
    int y = 0;
    y =(a | b) - (a & b);
    return(y);
}

void stampa(char bu[], int n, int m) {
    char buf[100], bb[100]; int k;
    k = m - 1; buf[m] = 0;
    for(k = m - 1; k >= 0; k--) {
        if((n & 1) > 0) buf[k] = 49;
        else buf[k] = 48;
        n = int(n / 2);
    }
    bb[0] = 0; aggiu(bb, bu); aggiu(bb, buf);
    cout << bb << endl; k = len(bb);
    bb[k] = 13; bb[k + 1] = 10;
    fwrite(bb, k + 2, sizeof(char), sto);
}

int len(char s[]) {
    int l = 0;
    while(s[l] != 0) l++;
    return(l);
}

void aggiu(char buf[], char buf1[]) {
    int i, j, j1;
    i = len(buf); j = 0; j1 = 0;
l1: buf[i + j] = buf1[j1];
    if(buf[i + j] == 10) {j1++; goto l2;}
    if(buf[i + j] == 13) {j1++; goto l2;}
    if(buf[i + j] != 0) {j++; j1++; goto l1;}
l2: buf[i + j] = 0;
}

void ouconv(char buf[], int v, int nc) {
    int j, va, c, k;
    k = len(buf); va = vabs(v);
    for(j = 0; j < nc; j++) buf[ j + k] = 32;
    c = va % 10;
```

```
    for(j = 1; j <= nc; j++) {
        buf[k + nc - j] = c + 48;
        va = va / 10; c = va % 10;
        if(va == 0) break;
    }
    if(v < 0) buf[k + nc - j - 1] = 45;
    buf[k + nc] = 0;
    return;
}
```

A.2 Routine to find a parity matrix for a two errors correction code

The following routine, written in C++ language, finds a parity matrix for a two bit error correction code using the rules described in Chap. 7, Sect. 7.2.2. The matrix depends on the length of data bits. All the results are written on a file. Also in this case the routine is composed of different subroutines.

INPUTS

- the number of data bytes (write 0 if you want to finish);
- the number of matrices that we want to find;
- the maximum number of searches we want to do.

OUTPUTS

- the number of data bits;
- the number of check bits;
- the maximum number of 1s per row;
- For all the matrices found: (a) the number of 1s per for column; (b) the number of 1s per row;
- how many searches it made;
- the number of matrices found;
- the number of non-equivalent matrices found.

ROUTINE

```
#include <iostream.h>
#include <stdio.h>
#include <stdlib.h>
#include <malloc.h>

/* Service routines */
int add(int a, int b);
int fnbit(int a);
int len(char s[]);
void ouconv(char buf1[], int v,int nc);
```

```
void scrivo(char bu[]);
void stampa(char bu[], int n, int m);
void stampa(int n1, int m, int a[]);
void aggiu(char a[], char b[]);
int testset(int a);
void clear(int a);
void hpsrt(int n1, int c[]);

int gencod(int n, int cod[]);

FILE *sto;
int *bit, *tcan, p2[32], px[32], nbit, min2, cod[65536];
/* *bit tells us where the table is,
   cod is the code that we find */
int pilcod[1048576], lista[1048576], codici[1048576];
char buf[80];

int main() {
    int i,nb;
    /* generation of the powers of 2 */
    p2[0] = 1; px[0] = ~0;
    for(i = 1; i < 32; i++) {
        p2[i] = p2[i - 1] << 1;
        px[i] = px[i - 1] & (~(1 << (i - 1)));
    }
    sto = fopen("result","wb");
    if(sto == 0) {
        buf[0] = 0; aggiu(buf, "error in open file result");
        scrivo(buf); return(0);
    }
    i = 65536; bit = (int *)malloc(i); cout << "malloc " << i <<
endl;
    /* we set a space in the memory to save the big table */
l1: cout << "number of data byte [0 end] "; cin >> nb; if(nb <= 0)
goto l2;
    /* we call the routine for the generation of the code */
    gencod(nb, cod);  goto l1;
l2: fclose(sto); return(0);
}

int gencod(int nb,int cod[]) {
    int ier, i, j, kk, kms = 4, ktot, n, t;
    int te, te1, tt, t1, t2, ng, k, space;
    int nbyte, ngen, ncan, nmax;
    /* kms is the maximum number of 1s per row;
       ng is the number of codewords that we found;
       min2 is the number of rows of the identity matrix;
       nbit is the number of the codewords we need */
    int tent, nlista, maxp, maxpl, minpl, minp;
    int tpesi[512][32], maxtent;
    /* nlista is the number of set of marked positions */
    int ssss[512];
    char buf[80];
    nbit = nb * 8; /* we find the number of bits */
    tcan = (int *)malloc(1907714 * 4);
    if(tcan == 0) return(0);
    cout << "max number of matrix "; cin >> nmax;
    /* it asks how many matrices we want to find */
```

```
      if(nb < 3) kms = 4;
      if(nb >= 4) kms = 6;
      if(nb >= 8) kms = 7;
      if(nb >= 16) kms = 10;
      minp = 1; maxp = 1024;
      cout << "max number of searches "; cin >> maxtent;
      /* it asks how many searches it has to do */
171:
      n = 0; t = nb; /* we try with another kms */
11:
      if(t > 1) {n++; t = t >> 1; goto 11;}
      tt = 1; min2 = 0;
12: t1 = nbit + min2; /* number of the rows we need */
      t2 = 1 + t1 + t1 * (t1 - 1) / 2;
    /* a factor for the Hamming inequality */
      if(tt < t2) {min2++; tt = tt << 1; goto 12;}
         /* tt is the maximum number of bits */
         /* we begin the search from the number of check bits
         necessary for BCH codes and so on */
      if(nb >= 1) {min2++; tt = tt << 1;}
      if(nb >= 4) {min2++; tt = tt << 1;}
      if(nb >= 32) {min2++; tt = tt << 1;}
      ngen = 0; ncan = 0; tent = 0;
      nbyte = tt / 8; space = nbyte / 4; bit = (int *)malloc(nbyte);
      if(bit == 0){
         buf[0] = 0; aggiu(buf, "out of memory");
         scrivo(buf); return(0);
      }
      for(i = 0; i < 1907713; i++) tcan[i] = 0;
      for(i = 0; i < space; i++) bit[i] = 0;
      /* we put all 0 row in the table */
      t = 1; ng = min2 + 1;
         /* we begin the search; after the identity matrix
         there is a vector with all 1 in the last 4 positions */
      for(i = 0; i < min2; i++) {
            /* we put the identity in the table */
            cod[i] = t; j = testset(t);
            tpesi[min2][i] = 0; t = t << 1;
      }
      ktot = 0; codici[0] = 0; codici[1] = 0;
      /* we fix the first 2 rows */
      for(i = 4; i <= kms; i++) {
         for(k = 0; k < tt; k++) {
            if(fnbit(k) == i) {
               /* There are already 2 codewords in the matrix */
               if(fnbit(add(k, codici[0])) < 3) goto 182;
               if(fnbit(add(k, codici[1])) < 3) goto 182;
               codici[ktot] = k; ktot++;
               /* we must have more than 3 1s per row,
               ktot is the maximum number in the generated table */
            }
182:         n = n;
         }
      }
      buf[0] = 0; aggiu(buf, "number of bit ");
      ouconv(buf, nbit, 4); scrivo(buf);
      buf[0] = 0; aggiu(buf, "number of bit code ");
      ouconv(buf, min2, 4); scrivo(buf);
```

```
    buf[0] = 0; aggiu(buf, "number max bit 1 ");
    ouconv(buf, kms, 4); scrivo(buf);
    pilcod[min2] = 0; pilcod[ng] = 1;
    /* we take the first 2 codewords */
    cod[min2] = codici[0];
    for(i = 0; i < 4; i++) tpesi[min2][i] = 1;
    i = testset(codici[0]);
    for(i = 0; i < tt; i++) {
        j = fnbit(i);
        if(j == 2) j = testset(i);
        /* we mark in the table the row that we cannot choose */
    }
15:
    /* ng is the position that it chose */
    if(pilcod[ng] >= ktot) goto 16;
    nlista = 1; k = codici[pilcod[ng]]; lista[0] = k;
    if(testset(k) == 0) {pilcod[ng]++; goto 15;}
    /* if it was already marked we choose another row */
    for(i = 0; i < ng; i++) {
        kk = add(cod[i], k);
        if(testset(kk) == 0) {
        /* if the sum was already marked, we choose another vector
        and we clear all the  information concerning this row */
        for(j = 0; j < nlista; j++)
            clear(lista[j]);
        pilcod[ng]++; goto 15;
        }
        lista[nlista] = kk; nlista++;
    }
    cod[ng] = k;
    for(j = min2 - 1; j >= 0; j--) {
        /* we find the number of 1s for every row of the identity ma-
trix */
        tpesi[ng][j] = tpesi[ng - 1][j]; if((k & (1 << j)) > 0)
tpesi[ng][j]++;
        if(tpesi[ng][j] < (minp - (t1 - ng))) {
            /* even if it chooses the all 1s vector it doesn't reach
the minimum */
            for(j = 0; j < nlista; j++)
                clear(lista[j]);
            pilcod[ng]++;  goto 15;
        }
        if(tpesi[ng][j] > maxp) {
            /* even if it chooses the all 0 vector the number of 1s
is bigger than
        the maximum */
            for(j = 0; j < nlista; j++)
                clear(lista[j]);
            pilcod[ng]++; goto 15;
        }
    }
    if(nbit > 1 + ng - min2) {
        /* we haven't found the minimum number of codewords */
        ng++; pilcod[ng] = pilcod[ng - 1] + 1; goto 15;
    }
    ngen++; maxpl = 0; minpl = 10000;
    for(i = 0; i < min2; i++) {
        /* if(tpesi[ng][i] < minp) goto 177;
```

```
    if(tpesi[ng][i] > maxp) goto 177; */
    if(tpesi[ng][i] > maxpl) maxpl = tpesi[ng][i];
    /* if the number of 1s per row is bigger than a maximum
     this is the new maximum */
    if(tpesi[ng][i] < minpl) minpl = tpesi[ng][i];
    /* if the number of 1s per row is bigger than a minimum
     this is the new minimum */
}
ier = 0; if(maxpl < maxp ) { maxp = maxpl; ier = 1;}
if(minpl > minp) {minp = minpl; ier = 1;}
ier = 0; if(maxpl <= maxp ) {maxp = maxpl; ier=1;}
if(minpl >= minp) {minp = minpl; ier = 1;}
buf[0] = 0;
for(i = 0; i < min2; i++)
    ouconv(buf, tpesi[ng][min2 - i - 1], 4); aggiu(buf," kms =");
ouconv(buf, kms, 4); scrivo(buf);
if(ier == 0) goto 177;
/* we find the number of 1s  of the whole matrix */
te = 1; te1 = 1;
for(j = 0; j < min2; j++) {
    te =(te * tpesi[ng][min2 - j - 1]) % 1907713;
    te1 = te1 * tpesi[ng][j] % 1970161;
}
/* we find the remainder of a division with 2 prime numbers, with
the first division we find the address to put the vector and with the
other division we find the value to put */
    if(tcan[te] == 0) {tcan[te] = te1; goto 178;}
    if(tcan[te] == te1) goto 177;
178:
    ncan++; buf[0] = 0; aggiu(buf, "solution  byte =");
    ouconv(buf, nb, 5); aggiu(buf, " cod =");
    ouconv(buf, min2, 5);aggiu(buf, " matrix ");
    ouconv(buf, ngen,10);
    aggiu(buf," non equivalent "); ouconv(buf,ncan,10); scrivo(buf);
    /* we write the non-equivalent matrices */
    for(i = min2; i <= ng; i++) ssss[i - min2] = cod[i];
    hpsrt(ng - min2 + 1, ssss); /* we set the matrices in a heap sort
setting */
    stampa(ng - min2, min2, ssss);
    goto 166;
177:
    if(ngen > nmax) {free(bit); free(tcan); return(0);}
    /* we found more matrices and we do not write them */
16:
    tent++;
    if(tent > maxtent) {
        buf[0] = 0; aggiu(buf, "end search ");
        ouconv(buf, maxtent, 10); scrivo(buf);
        /* we tried all the searches */
        buf[0] = 0; aggiu(buf, "number of matrix");
        ouconv(buf, ngen, 10);
        aggiu(buf,"  number of non equivalent matrix ");
        ouconv(buf, ncan, 10);
        scrivo(buf); free(bit); free(tcan); return(0);
    }
    ng--; /* we come back to the beginning with one more check bit */
    if(ng <= min2+2) { /* we go to the already fixed vectors */
        if(ncan > 0) { /* There are some equivalent matrices */
```

```
              buf[0]=0; aggiu(buf, "number of matrices");
              ouconv(buf,ngen,10);
              aggiu(buf, "  number of non equivalent matrices ");
              ouconv(buf,ncan,10);
              scrivo(buf); free(bit); free(tcan); return(0);
          }
        buf[0] = 0; aggiu(buf, "solution not found; max bit =");
      ouconv(buf, kms, 5);
      aggiu(buf, " check bits ="); ouconv(buf, min2, 3);
      aggiu(buf, " bit = "); ouconv(buf, nbit, 5);
        scrivo(buf); kms++;
        if(kms >= min2) {min2++; kms = 4;}
        /* we begin another search with one more check bit */
        free(bit); goto 171;
    }
166:
    k = cod[ng]; clear(k); /* we clear the big table */
    for(i = 0; i < ng; i++) {
        kk = add(cod[i], k); clear(kk);
    } pilcod[ng]++; goto 15;
}

int testset(int a) {
/* we test if the position is already written, if not we mark it */
    int lb, lbit, r;
    lb = a >> 5; lbit = a & 31;
    r = bit[lb] & (1 << lbit); if(r != 0) return(0);
    bit[lb] = bit[lb] | (1 << lbit); return(1);
}

void clear(int a) { /* we clear the table concerning a row */
    int lb, lbit;
    lb=a >> 5; lbit = a & 31; bit[lb] = bit[lb] & (~(1 << lbit));
}

int fnbit(int a) {
    int r = 0;
l1: if((a & 1) > 0) r++;
    a = a >> 1; if(a > 0) goto l1;
    return(r);
}

int add(int a,int b) {
    int r;
  r = (a | b) - (a & b); return(r);
}

int len(char s[]) {
    int l = 0;
    while(s[l] != 0) l++; return(l);
}

void aggiu(char buf2[], char buf1[]) {
    int i, j, j1; i = len(buf2); j = 0; j1 = 0;
l1: buf2[i + j] = buf1[j1];
    if(buf2[i + j] == 10) {j1++; goto l2;}
    if(buf2[i + j] == 13) {j1++; goto l2;}
    if(buf2[i + j] != 0) {j++; j1++; goto l1;}
```

```
12: buf2[i + j] = 0;
}

void ouconv(char buf1[], int v,int nc) {
    int j, va, c, k;
    k = len(buf1); va = abs(v);
    for(j = 0; j < nc; j++) buf1[j + k] = 32; c = va % 10;
    for(j = 1; j <= nc; j++) {
        buf1[k + nc - j] = c + 48;
        va = va / 10; c = va % 10;
        if(va == 0) break;
    }
    if(v < 0) buf1[k + nc - j - 1] = 45;
    buf1[k + nc] = 0;
    return;
}

void scrivo(char bu[]) {
    int l;
    l = len(bu);
    cout << bu << endl;
    bu[l] = 13; bu[l + 1] = 10; bu[l + 2] = 0;
    fwrite(bu, l + 2, sizeof(char), sto);
}

void stampa(int n, int m, int a[]) {
    int i, j;
    char bb[80], tbuf[10];
    tbuf[0] = 34; tbuf[1] = 0;
    bb[0] = 0; aggiu(bb, "constant P_MATRIX :P_MAT_128_10 := (");
    for(i = 0; i <= n; i++) {
        aggiu(bb, tbuf);
        for(j = 0; j < m; j++) {
            if((a[i] & (1 << (m - j - 1))) > 0) aggiu(bb, "1");
            else aggiu(bb, "0");
        }
        aggiu(bb, tbuf); if(i < n) aggiu(bb, ",");
        else aggiu(bb, ");");
        scrivo(bb);
        bb[0] = 0; aggiu(bb,"                                   ");
    }
}

void stampa(char bu[], int n1, int m) {
    char bub[1024], bb[1024];
    int k, n;
    k = m - 1; bub[m] = 0; n = n1;
    for(k = m - 1; k >= 0; k--) {
        if((n & 1) > 0) bub[k] = 49;
        else bub[k] = 48;
        n = n >> 1;
    }
    bb[0] = 0; aggiu(bb, bu);
    ouconv(bb, n1, 11); aggiu(bb, "  "); aggiu(bb,bub);
    cout << bb << endl; k = len(bb);
    bb[k] = 13; bb[k + 1] = 10;
    fwrite(bb, k + 2, sizeof(char), sto);
}
```

```
void hpsrt(int n1,int c[]) {
/*   setting with "heapsort"
     c vector with the data to set
     n1 number of datas to set */
     int k, m, l, np, n, l1, l2;
     n = n1; l1 = n / 2 - 1;
     for(np = 1; np < n1; np++) {
         for(l = l1; l >= 0; l--) {
             l2 = l; k = c[l2];
             while ((m = 2 * l2 + 1) < n) {
                 if(m < n - 1) {
                     if(c[m + 1] > c[m]) m = m + 1;
                 }
                 if(c[m] < k) break;
                 c[l2] = c[m]; l2 = m;
             }
             c[l2] = k;
         }
         l = c[n - 1]; c[n - 1] = c[0];
         c[0] = l; n = n - 1; l1=0;
     }
}
```

A.3 Routine to find a parity matrix to correct 2 erroneous cells

This program written in C++ finds a code corrector of two cells. All the results are written on a file. The search is made in a random way, so we need a generator, i.e. an odd number. The search is made with the constraints described in Chap. 7 (Sect. 7.3.1) and is composed by the identity matrix and by a set of rows linearly independent by group of four.

INPUTS

- the number of data bits (0 to finish);
- the number of check bits;
- the generator of the code (an odd number).

OUTPUTS

- it writes how many words are found with the chosen generator;
- it writes the code found.

ROUTINE

```
#include <iostream.h>
#include <stdio.h>
#include <stdlib.h>

/* Service routines */
int add(int a, int b);
int len(char s[]);
void ouconv(char buf1[], int v, int nc);
void scrivo(char bu[]);
void stampa(char bu[], int n, int m);
void stampa(int n1, int m, int a[]);
void aggiu(char a[], char b[]);
int gencod(int n, int min2, int gen);

FILE *sto;
int cod[65536];
int no[10000000];
char buf[200];

int main() {
    int i, nbit, min2, ntot, kmax, gen;
    sto = fopen("result", "wb");
    if(sto == 0) {
        buf[0] = 0; aggiu(buf, "error in open file result");
        scrivo(buf); return(0);
    }
    buf[0] = 0; aggiu(buf,"Program GENCODB.CXX"); scrivo(buf);
l1:
    cout << "number of data bits [0 end] ";
    cin >> nbit; if(nbit <= 0) goto l2;
/* it asks the number of data bits we want to consider */
    cout << "number of parity bits      ";
    cin >> min2; kmax = 1 << min2;
/* it asks the number of check bits we want to consider */
    buf[0] = 0; aggiu(buf, "number of bit      ");
    ouconv(buf, nbit, 5); scrivo(buf);
    buf[0]= 0; aggiu(buf, "number of bit code ");
    ouconv(buf, min2, 5); scrivo(buf);
    cout << "generator [1 or odd] "; cin >> gen;
/* it asks what generator (odd number) we want to consider */
l3:  i = gencod(nbit, min2, gen);
    /* it calls the routine to generate the code */
    if(i != 0) {
        buf[0] = 0; aggiu(buf, "error finds only ");
        ouconv(buf, i, 5);
        aggiu(buf, " elements with generator =");
        ouconv(buf, gen, 6);
        scrivo(buf); gen = gen + 2; if(gen < kmax) goto l3;
        /* it takes another generator: the next odd number */
        goto l1;
    }
    buf[0] = 0; aggiu(buf, "generator ");
    ouconv(buf, gen, 5); scrivo(buf);
    ntot = nbit + min2; stampa(ntot, min2, cod);
    goto l1;
l2:  fclose(sto); return(0);
```

```
}

int gencod(int nbit,int min2,int gen) {
    int i, j, j1, k, k1, k2;
    int k3, i1, kmax, ntot, nby, nbi, nc;
    kmax = 1 << min2;
    /* number of words that we can write with min2 bits */
    ntot = nbit + min2; k = kmax / 8; nc = 0;
    for(i = 0; i < k; i++) no[i] = 0;
    /* it puts a 0 in the forbidden-words table */
    k = 1;
    for(i = 0; i < min2; i++) { /* it puts the identity matrix */
        nby = k >> 3; nbi = k & 7;
        if((no[nby] & (1 << nbi)) > 0) return(-1);
        /* if the word is already a forbidden word we can't choose it
*/
        no[nby] = no[nby] | (1 << nbi); cod[i] = k;
        /* otherwise it chooses the word and marks the forbidden-
words table */
        for(j1 = 0; j1 < i; j1++) {
            k1 = add(k, cod[j1]); nby = k1 >> 3; nbi = k1 & 7;
            /* it sums the chosen word with the others in the code */
            if((no[nby] & (1 << nbi)) != 0) return(-1);
            /* if the sum is marked we can't choose the word */
            no[nby] = no[nby] | (1 << nbi);
            /* otherwise it chooses the word and marks the forbidden-
word table */
        }
        k = k << 1;
    }
    k = 0; j = min2;
    for(i = 0; i < nbit; i++) {
l1:         k = (k + gen) % kmax;
        /* it takes another word with distance g from the previous */
        nc++;if(nc >= kmax) return(i);
        /* if we can write the word with min2 bits we can choose it
*/
        nby = k >> 3; nbi = k & 7;
        if((no[nby] & (1 << nbi)) > 0) goto l1;
        /* if the word is already marked we can't choose it */
        for(i1 = 0; i1 < j; i1++) {
            k1 = add(k, cod[i1]); nby = k1 >> 3; nbi = k1 & 7;
            if((no[nby] & (1 << nbi)) > 0) goto l1;
            /* if the sum is already marked we can't choose the word
*/
            if((i1 & 1) == 1) {
                /* i1 is an odd number */
                k2 = add(k1, cod[i1 - 1]); nby = k2 >> 3;
                nbi = k2 & 7; if((no[nby] & (1 << nbi)) > 0) goto l1;
            }
            /* if the sum with the previous word is already marked
        we can't choose it */
        }
        if((j & 1) == 1) { /* if min2 is an odd number */
            k1 = add(k, cod[j - 1]); /* it sums the word with the row
in j-1 */
            for(i1 = 0; i1 < j - 1; i1++) {
                k2 = add(k1, cod[i1]); nby = k2 >> 3;
```

```
                    nbi = k2 & 7;
                    if((no[nby] & (1 << nbi)) > 0) goto l1;
                    /* if the sum with the other words in the code is al-
ready marked

                    we cannot choose the word */
                    if((i1 & 1) == 1) {
                        k3 = add(k2, cod[i1 - 1]);
                        nby = k3 >> 3; nbi = k3 & 7;
                        if((no[nby] & (1 << nbi)) > 0) goto l1;
                    }
                    /* if the sum of 2 cells is already marked we can't
choose the word */
                }
            }
        nby = k >> 3; nbi = k & 7; no[nby] = no[nby] | (1 << nbi);
        /* it marks the word in the forbidden-words table */
        for(i1 = 0; i1 < j; i1++) {
            k1 = add(k, cod[i1]); nby = k1 >> 3;
            nbi = k1 & 7; no[nby] = no[nby] | (1 << nbi);
            /* it marks the sum with other words in the forbidden-
words table */
            if((i1 & 1) == 1) { /* i1 is an odd number */
                k1 = add(k1, cod[i1 - 1]); nby = k1 >> 3;
                nbi = k1 & 7; no[nby] = no[nby] | (1 << nbi);
            } /* it marks the new sum in the table to correct a cell
and
                another error */
        }
        if((j & 1) == 1) { /* j is an odd number */
            k1 = add(k, cod[j - 1]); /* it sums with the previous
word */
            for(i1 = 0; i1 < j - 1; i1++) {
                k2 = add(k1, cod[i1]); nby = k2 >> 3; nbi = k2 & 7;
                no[nby] = no[nby] | (1 << nbi);
                /* it marks the sum with other words to correct a
word plus a cell */
                if((i1 & 1) == 1) {
                    k3 = add(k2, cod[i1 - 1]);
                    nby = k3 >> 3; nbi = k3 & 7;
                    no[nby] = no[nby] | (1 << nbi);
                }
                /* it marks the sum of 2 cells to correct 2 cells */
            }
        }
        cod[j]=k;j++; /* it chooses the word */
    }
    return(0);
}
```

Appendix B: BCH code

This Appendix is a collection of routines describing the encoding and the decoding of a classical BCH code. The BCH codes taken into examination are the primitive and narrow-sense ones. The routines implement the description of Chap. 8.

Some routines are written in Matlab language and some others in C language. The aim is to describe practical implementations of the code with all the details needed by the code.

A lot of these routines need some subroutines to perform multiplication, to find parameters and so on. Also these supporting routines are described here. For each routine there is a brief description of its functionality, its inputs and its outputs.

B.1 Routine to generate the BCH code parameters

In the following the implementation of a Matlab function able to find the generator polynomial and the minimal polynomials for a narrow-sense binary BCH code is described.

The function works well if the error correcting capability of the code is not very high. Also the length of the code can't be too high as Matlab is not able to guarantee computations for numbers bigger then 2^{15}.

INPUTS

- dim: the dimension of the field $GF(2^m)$ in which the code exists, e.g. if the code is in $GF(2^{13})$ then $dim=13$.
- t: the error correcting capability of the code.

OUTPUTS

- $GENPOLY$: the coefficients of the generator polynomial of the code.
- $FACTORS$: the t minimal polynomials of the code.

ROUTINE

```
function [GENPOLY, FACTORS] = bch_prim(dim, t)
```

```
m = gfprimdf(dim);
cs = gfcosets(dim);
pl = gfminpol(cs(2 : t + 1, 1), m);
FACTORS = pl;
GENPOLY = gftrunc(pl(1, :));
for i = 2 : t
    GENPOLY = gfconv(GENPOLY, gftrunc(pl(i, :)));
end;
return;
```

B.2 Routine to encode a message

In the following a C routine encodes a message using a BCH code. The encoding
is implemented trying to simulate the hardware machine as described in Chap. 8
(Sect. 8.2.1). Even if it is not an output of this routine, it must be underlined that
the written message is generated at this phase. This message contains in the first
PAR positions the parity vector and in the last *BCH_K* positions the data message.

INPUTS

- *BCH_K*: the length of data vector to be encoded.
- *PAR*: the length of parity vector that the chosen BCH code needs.
- *genpoly*: the coefficients of the generator polynomial of the code. These coefficients can be found with the previous Matlab routine.

OUTPUTS

- *PARITA*: generated parity vector.

ROUTINE

```
void encode(void)
{
    int i, j;
    int rientro;

    for (i = 0; i < PAR; i++)
        parita[i] = 0;
    for (i = BCH_K - 1; i >= 0; i--) {
        rientro = dato[i] ^ parita[PAR - 1];
        for (j = PAR - 1; j > 0; j--)
            parita[j] = parita[j - 1] ^ (rientro & genpoly[j]);
        parita[0] = rientro;
    }
}
```

B.3 Routine to calculate the syndromes of a read message

In the following a C-routine executes the first step of the decoding algorithm, that is the computation of the syndromes. The calculation is performed serially as described in Chap. 8 (Sect. 8.3.1). In the following the 5-error case is considered. Observe that the first division gives only the pre-syndromes as output. These vectors have to be evaluated in α as described in Chap. 8. The powers of α considered are the first five, because we are implementing primitive narrow-sense BCH codes. Each evaluation is represented by a matrix as will be described in the next section.

INPUTS

- *DIM_CAMPO*: the dimension of the code in which the code exists, that is if we are dealing with a code in $GF(2^{15})$, *DIM_CAMPO* will be 15.
- *polmin1, polmin2, polmin3, polmin4, polmin5*: the coefficients of the minimal polynomials of α, α^3, α^5, α^7 and α^9. These coefficients can be found with the Matlab routine described in Sect. B.1.

OUTPUTS

- *S1, PS3, PS5, PS7, PS9*: pre-syndromes vector for α, α^3, α^5, α^7 and α^9. Observe that in case of *S1*, the pre-syndrome is equivalent to the final syndrome.

ROUTINE1

```
void syndrome()
{
    int i, j;
    int rientro1, rientro2, rientro3, rientro4, rientro5;

    for (i = 0; i < DIM_CAMPO; i++) {
        S1[i] = 0;
        PS3[i] = 0;
        PS5[i] = 0;
        PS7[i] = 0;
        PS9[i] = 0;
    }
    for (i = BCH_N - 1; i >= 0; i--) {
        rientro1 = S1[DIM_CAMPO - 1];
        rientro2 = PS3[DIM_CAMPO - 1];
        rientro3 = PS5[DIM_CAMPO - 1];
        rientro4 = PS7[DIM_CAMPO - 1];
        rientro5 = PS9[DIM_CAMPO - 1];
        for (j = DIM_CAMPO - 1; j > 0; j--) {
            S1[j] = S1[j - 1] ^ (rientro1 & polmin1[j]);
            PS3[j] = PS3[j - 1] ^ (rientro2 & polmin2[j]);
```

```
                PS5[j] = PS5[j - 1] ^ (rientro3 & polmin3[j]);
                PS7[j] = PS7[j-1] ^ (rientro4 & polmin4[j]);
                PS9[j]=PS9[j-1] ^ (rientro5 & polmin5[j]);
            }
            S1[0]=rientro1 ^ ricevuto[i];
            PS3[0]=rientro2 ^ ricevuto[i];
            PS5[0]=rientro3 ^ ricevuto[i];
            PS7[0]=rientro4 ^ ricevuto[i];
            PS9[0]=rientro5 ^ ricevuto[i];
        }
    }
```

INPUTS

- *S1*, *PS3*, *PS5*, *PS7*, *PS9*: pre-syndromes vector for α, α^3, α^5, α^7 and α^9 from previous routine.
- *al3*, *al5*, *al7*, *al9*: these are four matrices of dimension *DIM_CAMPO* × *DIM_CAMPO* that describe how the scrambling of the bits must be performed in order to have the correct evaluation.

OUTPUTS

- *S1*, *S3*, *S5*, *S7*, *S9*: five syndromes, each of length *DIM_CAMPO*.

ROUTINE2

```
void valutazione_sindromi(void)
{
    int i, j;

    for (i = 0; i < DIM_CAMPO; i++) {
        S3[i] = 0;
        S5[i] = 0;
        S7[i] = 0;
        S9[i] = 0;
    }
    for (i = 0; i < DIM_CAMPO; i++) {
        for(j = 0; j < DIM_CAMPO; j++) {
            S9[i] = S9[i] ^ (al9[i][j] & PS9[j]);
            S7[i] = S7[i] ^ (al7[i][j] & PS7[j]);
            S5[i] = S5[i] ^ (al5[i][j] & PS5[j]);
            S3[i] = S3[i] ^ (al3[i][j] & PS3[j]);
        }
    }
}
```

B.4 Routine to calculate the evaluation matrices

As described in the previous section, we need to have the matrices of dimension $DIM\_CAMPO \times DIM\_CAMPO$ to perform the evaluations necessary to the syndromes computation.

In the following a Matlab routine performs the computation necessary to have the evaluation matrices. The routine contains two nested routines and is written using the communication Matlab toolbox, with the aid of the symbolic math.

INPUTS

- T: it is a matrix that represents all the vectors of the field $GF(2^m)$ so that it has dimension $(2^m-1) \times m$.
- a: it represents a vector of symbols in $GF(2^m)$. Suppose for example that we are searching a vector in $GF(2^4)$. First of all we have to define 4 symbols with the command

$$\text{syms } A\ B\ C\ D\ real$$

  then the vector a is defined as $a-[A\ B\ C\ D]$.
- pow: in the routine it is not represented as an input but it is easy changeable. The variable is declared at the very beginning and it represents the dimension of the field we are working on, that is it is the same as DIM_CAMPO of the previous routines.
- $div1$, $div2$, $div3$, $div4$: they are placed at the beginning of the routine as the variable pow. They represent the α power for which we compute the matrices.

OUTPUTS

- $A3$, $A5$, $A7$, $A9$: five matrices to perform the evaluations, each one of dimension $pow \times pow$.

ROUTINE

```
function [A3, A5, A7, A9] = valutazione(a, T)

global pow

pow = 15;
div1 = 3;
div2 = 5;
div3 = 7;
div4 = 9;

P3 = valutazione(div1, a, T)
A3 = matrice(P3, a)
P5 = valutazione(div2, a, T)
A5 = matrice(P5, a)
```

```
P7 = valutazione(div3, a, T)
A7 = matrice(P7, a)
P9 = valutazione(div4, a, T)
A9 = matrice(P9, a)

function P = valutazione(div, a, T)
global pow T
P = a;
D = a;
for i = 2 :(pow)
    P(i) = 0;
end;
for i = 2 : (pow)
    z = (div * (i - 1)) + 1;
    P = P + D(i) * (T(z, :));
end;
return;


function A = matrice(P, a)
global pow
A = zeros(pow);
for i = 1 : pow
    B = char(P(i));
    for z = 1 : pow
        j = a(z);
        j = char(j);
        C = strfind(B, j);
        if C > 0
                A(i, z) = 1;
          end;
          z = z + 1;
      end;
      i = i + 1;
end;
return;
```

B.5 Routines to compute operations in a Galois field

The following step of the decoding routine is the calculation of the coefficients of the error locator polynomial. The lambda coefficients are vectors of $GF(2^m)$ and are computed starting from the syndromes that are also vectors in $GF(2^m)$.

In order to perform the necessary operations we describe the implementation of the Galois field operations in the following.

All the operations are described according to the theory depicted in Chap. 6.

B.5.1 Sum in a Galois field

The sum in a Galois field $GF(2^m)$ is only the exclusive-OR of all the operand bits.

INPUTS

- *DIM_CAMPO*: the dimension of the working field. If, for example, we are working on $GF(2^{15})$, *DIM_CAMPO* is 15.
- *a*, *b*: two vectors of length *DIM_CAMPO* to be summed.

OUTPUTS

- *ipSum*: vector that contains the sum of length *DIM_CAMPO*.

ROUTINE

```
void somma(int a[DIM_CAMPO], int b[DIM_CAMPO], int ipSum[DIM_CAMPO])
{
   int i,j;

   for(i = 0; i < DIM_CAMPO; i++)
       ipSum[i] = a[i] ^ b[i];
}
```

B.5.2 Multiplication in a Galois field

The multiplication in a Galois field $GF(2^m)$ is a more complicated operation as already described in Chap. 6.

INPUTS

- *DIM_CAMPO*: the dimension of the working field. If, for example, we are working on $GF(2^{15})$, *DIM_CAMPO* is 15.
- *a*, *b*: two vectors of length *DIM_CAMPO* to be multiplied.
- *matrix*: it is the association rule for the field we are working on. It is represented as a matrix of dimension $DIM\_CAMPO \times DIM\_CAMPO$.

OUTPUTS

- *ipProd*: vector that contains the result of length *DIM_CAMPO*.

ROUTINE

```
void moltiplicazione(int a[DIM_CAMPO], int b[DIM_CAMPO], int
ipProd[DIM_CAMPO])
{
    int tmp[ 2 * DIM_CAMPO - 1];
    int i, j, l, z;

    for(i = 0; i <= 2 * DIM_CAMPO - 2; i++)
        tmp[i] = 0;
    for(l = 0; l < DIM_CAMPO; l++) {
        for(j = 0; j <= l; j++)
        tmp[l] = tmp[l] ^ (a[j] & b[l - j]);
    }
    for(l = 0; l <= DIM_CAMPO - 2; l++) {
        z = DIM_CAMPO + l;
        for(j = l + 1; j < DIM_CAMPO; j++)
            tmp[z] = tmp[z] ^ (a[j] & b[z - j]);
    }
    for(j = 0; j < DIM_CAMPO - 1; j++) {
        for(i = 0; i < DIM_CAMPO; i++)
            tmp[i] = tmp[i] ^ (tmp[j + DIM_CAMPO] & matrix[j][i]);
    }
    for(i = 0; i < DIM_CAMPO; i++)
        ipProd[i] = tmp[i];
}
```

B.5.3 Division in a Galois field

In the following, the division in a Galois field $GF(2^m)$ is implemented as a series of multiplications, using the routine described in the previous section.

INPUTS

- *DIM_CAMPO*: the dimension of the working field. If, for example, we are working on $GF(2^{15})$, *DIM_CAMPO* is 15.
- *a, x*: the vectors of length *DIM_CAMPO* that have to be divided.

OUTPUTS

- *ipDiv*: vector that contain the result of length *DIM_CAMPO*.

ROUTINE

```
void divisione(int a[DIM_CAMPO], int x[DIM_CAMPO], int ip-
Div[DIM_CAMPO])
{
    int i, j;
    int prod[DIM_CAMPO], prodx[DIM_CAMPO];
```

```
moltiplicazione(a, a, prod);
for(i = 0; i < DIM_CAMPO - 1; i++)
{
    moltiplicazione(prod, x, prodx);
    moltiplicazione(prodx, prodx, prod);
}
for(i = 0; i < DIM_CAMPO; i++)
    ipDiv[i] = prod[i];
}
```

B.6 Routine to calculate the lambda coefficients

There are a lot of implementations for this step of the decoding algorithm. We present here the Berlekamp-Massey algorithm with and without inversion.

B.6.1 Berlekamp-Massey algorithm with inversion

The implementation of the Berlekamp-Massey algorithm needs some more supporting routines to initialize the registers and to make some basic operations on the registers. In the following these routines are separated to describe the inputs and the outputs for each of them.

This routine tests if a vector is an all 0s vector.

INPUTS

- *DIM_CAMPO*: the dimension of the working field. If, for example, we are working on $GF(2^{15})$, *DIM_CAMPO* is 15.
- *test[DIM_CAMPO]*: a vector of length *DIM_CAMPO*.

OUTPUTS

- *an integer*: 1 if the vector is an all 0s vector and 0 in all the other cases.

ROUTINE

```
int verifica(int test[DIM_CAMPO])
{
    int i;

    for(i = 0; i < DIM_CAMPO; i++) {
        if(test[i] == 1) {
            return 1;
```

```
        }
    }
    return 0;
}
```

This routine tests the degree of a polynomial *Lambda*.

INPUTS

- *err*: the error correcting capability of the code.
- *Lambda*: it represents the polynomial of which we want to know the degree. In the implementation it is seen as a matrix of dimension *err* × *DIM_CAMPO*.

OUTPUTS

- *deg*: the degree of the polynomial *Lambda*.

ROUTINE

```
void grado()
{
    int i;

    deg = -1;
    for(i = err - 1; i >= 0; i--) {
        if(verifica(Lambda[i]) == 1) {
            deg = i + 1;
            return;
        }
    }
    if(deg == 0)
        printf("\nError");
    return;
}
```

This routine shifts a register.

INPUTS

- *err*: the error correcting capability of the code.
- *DIM_CAMPO*: the dimension of the working field. If, for example, we are working on $GF(2^{15})$, *DIM_CAMPO* is 15.
- *T[DIM_CAMPO]*: the register that has to be shifted.

OUTPUTS

- *T[DIM_CAMPO]*: the register shifted.

ROUTINE

```
void shift()
{
    int i, j;
    for(j = err - 1; j >= 1; j--) {
        for(i = 0; i < DIM_CAMPO; i++)
            T[j][i] = T[j - 1][i];
    }
}
```

This routine initializes the registers in order to perform the Berlekamp-Massey algorithm.

INPUTS

- *S1,S3,S5,S7,S9*: the syndromes.
- *DIM_CAMPO*: the dimension of the working field. If, for example, we are working on $GF(2^{15})$, *DIM_CAMPO* is 15.
- *T[DIM_CAMPO]*: the register that has to be shifted.
- *k*: it is used to understand which register needs to be initialized.
- *num*: due to the fact that we are working with binary field, the number of cycles for the Berlekamp-Massey algorithm is halved. In this case, two shifts have to be performed for each step so that the subroutine is called twice: the first time with *num=0* and the second one with *num=1*.

OUTPUTS

- *T*: the initialized register.

ROUTINE

```
void primo(int k, int num)
{
    if(num==0) {
        switch(k) {
            case 0: moltiplicazione(S1, S1, T[0]); break;
            case 1: moltiplicazione(T[2], T[2], T[0]); break;
            case 2: moltiplicazione(T[3], T[3], T[0]); break;
            case 3: moltiplicazione(T[4], T[4], T[0]); break;
            default: printf("\nError"); break;
        }
    }
```

```
    else {
        int i;
        switch(k) {
            case 0: for(i = 0; i < DIM_CAMPO; i++)
                        T[0][i] = S3[i]; break;
            case 1: for(i = 0; i < DIM_CAMPO; i++)
                        T[0][i] = S5[i]; break;
            case 2: for(i = 0; i < DIM_CAMPO; i++)
                        T[0][i] = S7[i]; break;
            case 3: for(i = 0; i < DIM_CAMPO; i++)
                        T[0][i] = S9[i]; break;
            default: printf("\nError"); break;
        }
    }
    return;
}
```

Finally this is the real Berlekamp-Massey algorithm.

INPUTS

- $S1, S3, S5, S7, S9$: the syndromes.
- err: the error correcting capability of the code.
- DIM_CAMPO: the dimension of the working field. If, for example, we are working on $GF(2^{15})$, DIM_CAMPO is 15.

OUTPUTS

- $Lambda$: the coefficients of the error locator polynomial as a matrix of dimension err x DIM_CAMPO.

ROUTINE

```
void algo()
{
    int i, k, j, direzione;
    int R1[DIM_CAMPO], R2[DIM_CAMPO];

    if(verifica(S1)+
        verifica(S3)+
        verifica(S5)+
        verifica(S7)+
        verifica(S9) == 0)
        deg=0;
    else {
/* Initialization */
        for(i = 0; i < err; i++) {
            for(j = 0; j < DIM_CAMPO; j++) {
                T[i][j] = 0;
                Lambda[i][j] = 0;
                K[i][j] = 0;
            }
        }
```

```
        for(i = 0; i < DIM_CAMPO; i++) {
            lam1[i] = 0;
            lam2[i] = 0;
            lam3[i] = 0;
            lam4[i] = 0;
            lam5[i] = 0;
            discre[i] = 0;
            T[0][i] = S1[i];
        }
        K[0][0] = 1;
        for(k = 0; k < err; k++) {
/* Discrepancy calculation */
            for(i = 0; i < DIM_CAMPO; i++)
                discre[i] = T[0][i];
            for(i = 1; i < err; i++) {
                moltiplicazione(Lambda[i - 1], T[i], R1);
                somma(discre, R1, discre);
            }
            direzione = verifica(discre);
            grado();
/* Lambda and K calculation */
            for(i = err - 1; i > 1; i--) {
                moltiplicazione(K[i], discre, R1);
                if((direzione == 0) || (deg > k)) {
                    for(j = 0; j < DIM_CAMPO; j++)
                        K[i][j] = K[i - 2][j];
                }
                else {
                    if(verifica(Lambda[i - 2]) == 0) {
                        for(j = 0; j < DIM_CAMPO; j++)
                            K[i][j] = 0;
                    }
                    else {
                        divisione(Lambda[i - 2], discre,K[i]);
                    }
                }
                somma(R1, Lambda[i], Lambda[i]);
            }
            moltiplicazione(K[1], discre,R1);
            if((direzione == 0) || (deg > k)) {
                for(j = 0; j < DIM_CAMPO; j++)
                    K[1][j] = 0;
            }
            else {
                for(j = 0; j < DIM_CAMPO; j++)
                    R2[j] = 0;
                R2[0] = 1;
                divisione(R2, discre, K[1]);
            }
            somma(R1, Lambda[1], Lambda[1]);
            moltiplicazione(K[0], discre,R1);
            for(j = 0; j < DIM_CAMPO; j++)
                K[0][j] = 0;
            somma(R1, Lambda[0], Lambda[0]);
 /* two shift registers */
            if(k != err - 1) {
                shift();
                primo(k, 0);
```

```
                    shift();
                    primo(k, 1);
            }
        }
        grado();
    }
    for(j = 0; j < DIM_CAMPO; j++) {
        lam1[j] = Lambda[0][j];
        lam2[j] = Lambda[1][j];
        lam3[j] = Lambda[2][j];
        lam4[j] = Lambda[3][j];
        lam5[j] = Lambda[4][j];
    }
}
```

B.6.2 Berlekamp-Massey algorithm without inversion

Also in this case, the implementation of Berlekamp-Massey algorithm needs some more supporting routines to initialize the registers and to make some basic operations on registers. These routines are the same as those one described before and will not be described here.

INPUTS

- *S1*, *S3*, *S5*, *S7*, *S9*: the syndromes.
- *err*: the error correcting capability of the code.
- *DIM_CAMPO*: the dimension of the working field. If, for example, we are working on $GF(2^{15})$, *DIM_CAMPO* is 15.

OUTPUTS

- *Lambda*: the coefficients of the error locator polynomial as a matrix of dimension *err* × *DIM_CAMPO*.

ROUTINE

```
void algo() {
    int i, k, j, direzione;
    int R1[DIM_CAMPO], R2[DIM_CAMPO];

    if(sindro0() == 0)
        deg = 0;
    else {
/* Initialization */
        for(i = 0; i < err + 1; i++) {
            for(j = 0; j < DIM_CAMPO; j++) {
                T[i][j] = 0;
                Lambda[i][j] = 0;
                K[i][j]= 0;
```

```
            }
        }
        for(i = 0; i < DIM_CAMPO; i++) {
            lam0[i] = 0;
            lam1[i] = 0;
            lam2[i] = 0;
            lam3[i] = 0;
            lam4[i] = 0;
            lam5[i] = 0;
            discre[i] = 0;
            delta[i] = 0;
            T[0][i] = S1[i];
        }
        Lambda[0][0] = 1;
        K[0][0] = 1;
        delta[0] = 1;
        for(k = 0; k < err; k++) {
/* Discrepancy calculation */
            for(i = 0; i < DIM_CAMPO; i++)
                discre[i] = 0;
            for(i = 0; i < err + 1; i++) {
                moltiplicazione(Lambda[i], T[i], R1);
                somma(discre, R1, discre);
            }
            direzione = verifica(discre);
            grado();
/* Lambda and K calculation */
            for(i = err; i >= 2; i--) {
                moltiplicazione(K[i - 1], discre, R1);
                moltiplicazione(Lambda[i], delta, R2);
                if((direzione==0) || (deg > k)) {
                    for(j = 0; j < DIM_CAMPO; j++)
                        K[i][j] = K[i - 2][j];
                }
                else {
                    for(j = 0; j < DIM_CAMPO; j++)
                        K[i][j] = Lambda[i - 1][j];
                }
                somma(R1, R2, Lambda[i]);
            }
            moltiplicazione(K[0], discre, R1);
            moltiplicazione(Lambda[1], delta, R2);
            if((direzione == 0) || (deg > k)) {
                for(j = 0; j < DIM_CAMPO; j++)
                    K[1][j] = 0;
            }
            else {
                for(j = 0; j < DIM_CAMPO; j++)
                    K[1][j] = Lambda[0][j];
            }
            somma(R1, R2, Lambda[1]);
            for(j = 0; j < DIM_CAMPO; j++)
                K[0][j] = 0;
            moltiplicazione(delta, Lambda[0], Lambda[0]);
            if((direzione != 0) & (deg<=k)) {
                for(j = 0; j < DIM_CAMPO; j++)
                    delta[j] = discre[j];
            }
```

```
/* two shift registers */
            if(k != err-1) {
                shift();
                primo(k, 0);
                shift();
                primo(k, 1);
            }
        }
        grado();
    }
    for(j = 0; j < DIM_CAMPO; j++) {
        lam0[j] = Lambda[0][j];
        lam1[j] = Lambda[1][j];
        lam2[j] = Lambda[2][j];
        lam3[j] = Lambda[3][j];
        lam4[j] = Lambda[4][j];
        lam5[j] = Lambda[5][j];
    }
}
```

B.7 Routine to execute Chien algorithm

An implementation of a Chien routine is now described. This routine is the last step of the decoding algorithm and it finds the roots of the error locator polynomial *Lambda* found in the previous section.

Before going through this routine we underline that we are using shortened BCH codes. For this reason, as said in Chap. 8, we need to initialize the Lambda coefficients of the Chien machine.

This routine initializes the *Lambda* coefficients for the length *BCH_K* of the code. The goal of the routine is very easy to understand: we find the coefficient suitable for the chosen length and we multiply this coefficient once for the first degree *Lambda* coefficient, twice for the second *Lambda* coefficient and so on.

INPUTS

- *lam1, lam2, lam3, lam4, lam5*: the five Lambda coefficients found by the Berlekamp-Massey algorithm.
- *al16384*: the multiplicative coefficient needed to initialize the coefficients. It is found in the way described in Chap. 8.
- *DIM_CAMPO*: the dimension of the working field. If, for example, we are working on $GF(2^{15})$, *DIM_CAMPO* is 15.

OUTPUTS

- *lam1, lam2, lam3, lam4, lam5*: the five initialized *Lambda* coefficients.

ROUTINE

```
void lam16384() {
    int i, supp1[DIM_CAMPO];

    moltiplicazione(lam1, al16384, supp1);
    for(i = 0; i < DIM_CAMPO; i++)
        lam1[i] = supp1[i];
    moltiplicazione(lam2, al16384, supp1);
    moltiplicazione(supp1, al16384, lam2);
    moltiplicazione(lam3, al16384, supp1);
    moltiplicazione(supp1, al16384, lam3);
    moltiplicazione(lam3, al16384, supp1);
    for(i = 0; i < DIM_CAMPO; i++)
        lam3[i] = supp1[i];
    moltiplicazione(lam4, al16384, supp1);
    moltiplicazione(supp1, al16384, lam4);
    moltiplicazione(lam4, al16384, supp1);
    moltiplicazione(supp1, al16384, lam4);
    moltiplicazione(lam5, al16384, supp1);
    moltiplicazione(supp1, al16384, lam5);
    moltiplicazione(lam5, al16384, supp1);
    moltiplicazione(supp1, al16384, lam5);
    moltiplicazione(lam5, al16384, supp1);
    for(i = 0; i < DIM_CAMPO; i++)
        lam5[i] = supp1[i];
}
```

Now that all the coefficients are initialized in the correct way, the Chien algorithm can begin its search. This algorithm needs the multiplication by α in the field.

INPUTS

- *DIM_CAMPO*: the dimension of the working field. If, for example, we are working on *GF(2^{15})*, *DIM_CAMPO* is 15.
- *lam*: the vector of length *DIM_CAMPO* that has to be multiplied.
- *alp*: the matrix that describes the multiplication rule in the field.

OUTPUTS

- *lam*: the vector multiplied.

ROUTINE

```
void moltiplicazione_alpha(int lam[DIM_CAMPO]) {
    int i, j;
    int lamtmp[DIM_CAMPO];

    for (i = 0; i < DIM_CAMPO; i++)
```

```
    lamtmp[i] = 0;
    for (i = 0; i < DIM_CAMPO; i++) {
        for(j = 0; j < DIM_CAMPO; j++)
        lamtmp[i] = lamtmp[i] ^ (alp[i][j] & lam[j]);
    }
    for (i = 0; i < DIM_CAMPO; i++)
    lam[i] = lamtmp[i];
}
```

These others routines, used in the Chien algorithm, are here reported without description, due to the simplicity of understanding.

ROUTINE that multiplies one element x alpha^2 in the field GF(2^15)

```
void moltiplicazione_alpha2(void)
{
    moltiplicazione_alpha(lam2);
    moltiplicazione_alpha(lam2);
}
```

ROUTINE that multiplies one element x alpha^3 in the field GF(2^15)

```
void moltiplicazione_alpha3(void)
{
    moltiplicazione_alpha(lam3);
    moltiplicazione_alpha(lam3);
    moltiplicazione_alpha(lam3);
}
```

ROUTINE that multiplies one element x alpha^4 in the field GF(2^15)

```
void moltiplicazione_alpha4(void)
{
    moltiplicazione_alpha(lam4);
    moltiplicazione_alpha(lam4);
    moltiplicazione_alpha(lam4);
    moltiplicazione_alpha(lam4);
}
```

ROUTINE that multiplies one element x alpha^5 in the field GF(2^15)

```
void moltiplicazione_alpha5(void)
{
    moltiplicazione_alpha(lam5);
    moltiplicazione_alpha(lam5);
    moltiplicazione_alpha(lam5);
    moltiplicazione_alpha(lam5);
    moltiplicazione_alpha(lam5);
}
```

Now the real routine composed of all these subroutines is very easy to implement. Observe that there is a check if the number of errors found is the same as the polynomial degree. If this check is not verified we have a decoding failure, visible as an all 1s vector.

INPUTS

- *DIM_CAMPO*: the dimension of the working field. If, for example, we are working on $GF(2^{15})$, *DIM_CAMPO* is 15.
- *BCH_N*: the codeword length for the chosen code.

OUTPUTS

- *errorfound*: the error vector. It is a vector of length *BCH_N* with a 1 in the erroneous positions. In order to have the corrected message, this vector must be added to the read message.

ROUTINE

```
void ricerca_errori(void)
{
    int erro, sum, posizione;
    int i, j;

    numerr = 0;
    for(j = 0; j < BCH_N; j++) {
        erro = 1 ^ lam1[0] ^ lam2[0] ^ lam3[0] ^ lam4[0] ^ lam5[0];
        posizione = 0;
        if(erro == 0) {
            for(i = 1; i < DIM_CAMPO; i++) {
                sum = lam1[i] ^
                      lam2[i] ^
                      lam3[i] ^
                      lam4[i] ^
                      lam5[i];
                if(sum == 0) {
                    posizione = 1;
                }
                else {
                    posizione = 0;
                    i = DIM_CAMPO;
                }
            }
        }
        errorfound[BCH_N - 1 - j] = posizione;
        if(errorfound[BCH_N - 1 - j] == 1) {
            numerr++;
        }
        moltiplicazione_alpha(lam1);
        moltiplicazione_alpha2();
        moltiplicazione_alpha3();
```

```
            moltiplicazione_alpha4();
            moltiplicazione_alpha5();
    }
    if(deg != numerr) {
        for(j = 0; j < BCH_N; j++)
            errorfound[j] = 1;
    }
    printf("Degree: %d   >>>>   Numerr: %d\n", deg, numerr);
}
```

B.8 Routine to find the matrix to execute the multiplication by alpha

Also the multiplication by α can be described by a matrix of dimension *DIM_CAMP* × *DIM_CAMPO*. The multiplication indeed is equivalent to a linear shift of the columns. The only column that needs some more computations is the last one, because its shift must be in the field.

This routine, written in Matlab language, finds this matrix necessary for the Chien algorithm.

As already described for the routine that finds the evaluation matrices, also the following routine uses the Matlab communication toolbox of symbolic math.

INPUTS

- *T*: it is a matrix that represents all the vectors of the field $GF(2^m)$ so that it has dimension $(2^m-1) \times m$.
- *a*: it represents a vector of symbols in $GF(2^m)$. Suppose for example that we are searching a vector in $GF(2^4)$. First of all we have to define 4 symbols with the command

  syms A B C D real

  then the vector *a* is defined as *a=[A B C D]*.
- *pow*: in the routine it is not represented as an input but it is easily changeable. The variable is declared at the very beginning and it represents the dimension of the field we are working on: it is the same as *DIM_CAMPO* of the previous routines.

OUTPUTS

- *Aalpha*: the matrix to be used in the multiplication of dimension *pow* × *pow*.

ROUTINE

```
function Aalpha = peralpha(a, T)

global pow

pow = 15;

Palpha = [sym(0)];
Palpha(1 : pow + 1) = [0 a];
Z = Palpha(pow + 1) * T(pow + 1, :);
Palpha = Palpha(1 : pow);
Palpha = Palpha + Z
Aalpha = matrice(Palpha, a)

function A = matrice(P, a, T)
global pow
A = zeros(pow);
for i = 1 : pow
    B = char(P(i));
    for z = 1 : pow
        j = a(z);
        j = char(j);
        C = strfind(B,j);
        if C > 0
            A(i, z) = 1;
        end;
        z = z + 1;
    end;
    i = i + 1;
end;
return
```

Appendix C: The Galois field $GF(2^4)$

In the following the representations of a Galois field $GF(2^4)$ and the tables to perform the sum and the multiplication are presented.

The data in the following table must be interpreted as in Table 2.5 of Chap. 2, i.e. the decimal form does not represent a number.

The primitive polynomial for this field is $p(x)=x^4+x+1$.

Table C.1. Field elements of $GF(16)$

Index form	Polynomial form	Binary form	Decimal form
0	0	0000	0
α^0	1	0001	1
α^1	α	0010	2
α^2	α^2	0100	4
α^3	α^3	1000	8
α^4	$\alpha+1$	0011	3
α^5	$\alpha^2+\alpha$	0110	6
α^6	$\alpha^3+\alpha^2$	1100	12
α^7	$\alpha^3+\alpha+1$	1011	11
α^8	α^2+1	0101	5
α^9	$\alpha^3+\alpha$	1010	10
α^{10}	$\alpha^2+\alpha+1$	0111	7
α^{11}	$\alpha^3+\alpha^2+\alpha$	1110	14
α^{12}	$\alpha^3+\alpha^2+\alpha+1$	1111	15
α^{13}	$\alpha^3+\alpha^2+1$	1101	13
α^{14}	α^3+1	1001	9

Table C.2. Additional table for *GF(16)*

	0	1	2	3	4	5	6	7	8	9	10	11	12	13	14	15
0	0	1	2	3	4	5	6	7	8	9	10	11	12	13	14	15
1	1	0	3	2	5	4	7	6	9	8	11	10	13	12	15	14
2	2	3	0	1	6	7	4	5	10	11	8	9	14	15	12	13
3	3	2	1	0	7	6	5	4	11	10	9	8	15	14	13	12
4	4	5	6	7	0	1	2	3	12	13	14	15	8	9	10	11
5	5	4	7	6	1	0	3	2	13	12	15	14	9	8	11	10
6	6	7	4	5	2	3	0	1	14	15	12	13	10	11	8	9
7	7	6	5	4	3	2	1	0	15	14	13	12	11	10	9	8
8	8	9	10	11	12	13	14	15	0	1	2	3	4	5	6	7
9	9	8	11	10	13	12	15	14	1	0	3	2	5	4	7	6
10	10	11	8	9	14	15	12	13	2	3	0	1	6	7	4	5
11	11	10	9	8	15	14	13	12	3	2	1	0	7	6	5	4
12	12	13	14	15	8	9	10	11	4	5	6	7	0	1	2	3
13	13	12	15	14	9	8	11	10	5	4	7	6	1	0	3	2
14	14	15	12	13	10	11	8	9	6	7	4	5	2	3	0	1
15	15	14	13	12	11	10	9	8	7	6	5	4	3	2	1	0

Table C.3. Multiplication table for *GF(16)*

	0	1	2	3	4	5	6	7	8	9	10	11	12	13	14	15
0	0	0	0	0	0	0	0	0	0	0	0	0	0	0	0	0
1	0	1	2	3	4	5	6	7	8	9	10	11	12	13	14	15
2	0	2	4	6	8	10	12	14	3	1	7	5	11	9	15	13
3	0	3	6	5	12	15	10	9	11	8	13	14	7	4	1	2
4	0	4	8	12	3	7	11	15	6	2	14	10	5	1	13	9
5	0	5	10	15	7	2	13	8	14	11	4	1	9	12	3	6
6	0	6	12	10	11	13	7	1	5	3	9	15	14	8	2	4
7	0	7	14	9	15	8	1	6	13	10	3	4	2	5	12	11
8	0	8	3	11	6	14	5	13	12	4	15	7	10	2	9	1
9	0	9	1	8	2	11	3	10	4	13	5	12	6	15	7	14
10	0	10	7	13	14	4	9	3	15	5	8	2	1	11	6	12
11	0	11	5	14	10	1	15	4	7	12	2	9	13	6	8	3
12	0	12	11	7	5	9	14	2	10	6	1	13	15	3	4	8
13	0	13	9	4	1	12	8	5	2	15	11	6	3	14	10	7
14	0	14	15	1	13	3	2	12	9	7	6	8	4	10	11	5
15	0	15	13	2	9	6	4	11	1	14	12	3	8	7	5	10

Appendix D: The parallel BCH code

This Appendix contains the Matlab routines used to get the matrices needed in the hardware implementation of the parallel BCH machines and the C-language code of the Berlekamp decoding, as described in Chap. 9. A global Matlab overview of BCH operations shows how these matrices could be used for the encoding and the decoding operations.

D.1 Routine to get the matrix for the encoding

The parallel encoding of the considered example needs a matrix of dimension 75×83. At every step of the encoding algorithm, the matrix is multiplied by a vector containing the data inputs of 8 bits and the 75 bits of the previous cycle. At the last step these 75 bits will be the remainder of the division between the message to be encoded (previously shifted) and the code generator polynomial. The result of the multiplication of the 83-bit vector by this matrix is a vector of 75 bits representing the register contents at every step.

INPUTS

- *dim*: the number of data bits encoded at the same time.
- *g*: the coefficients of the code generator polynomial.

OUTPUTS

- *PAR*: the matrix of dimension 75×83 used in the encoding algorithm.

ROUTINE

```
function PAR = get_PAR()
    dim = 8;
    g = [1 1 1 0 0 0 0 0 0 1 1 0 1 1 1 1 0 0 1 1 0 1 0 0 1 ...
         0 1 0 0 1 0 0 0 0 1 1 0 0 1 1 1 1 1 0 1 0 1 0 1 1 1 0 0 1 ...
         1 0 1 0 1 1 1 0 0 1 0 0 1 1 1 1 1 1 1 1 0];

    PAR = [[zeros(dim, 75 - dim), eye(dim)]', eye(75)]
    k = zeros(1, 75 + dim);
    for i = 1 : dim
        PAR = 1.* xor([k; PAR(1 : end - 1, :)], g' * PAR(end, :));
    end
end
```

D.2 Routine to get matrices for the syndromes

The parallel syndromes calculation can be implemented either as evaluations in α powers of the remainder of the division between the received message and the minimal polynomials either as evaluations in α powers of the received message (Sect. 9.5). For the first method we need the matrices to get the pre-syndromes and the matrices for their evaluation in α, α^3, α^5, α^7 and α^9 in order to have the final syndromes.

INPUTS

- *dim*: the number of data bits encoded at the same time.
- *g*: the coefficients of the field generator polynomial.
- *b1, b3, b5, b7, b9:* the coefficients of the 5 minimal polynomials of α, α^3, α^5, α^7, α^9.

OUTPUTS

- *B1, B3, B5, B7, B9*: the five matrices of dimension 15×23 used in the pre-syndromes computation.
- *A1, A3, A5, A7, A9*: the five matrices of dimension 15×23 used to evaluate the pre-syndromes in α, α^3, α^5, α^7 and α^9 in order to have the final syndromes.

ROUTINE

```
function [B1, B3, B5, B7, B9, A1, A3, A5, A7, A9] = get_SYN_REM()

    b1 = [1 1 0 0 0 0 0 0 0 0 0 0 0 0 0];
    b3 = [1 1 0 0 0 1 0 0 0 0 1 0 0 0 0];
    b5 = [1 1 0 1 0 0 0 0 0 0 0 1 0 0];
    b7 = [1 1 0 1 0 1 0 1 0 1 0 1 0 1 0];
    b9 = [1 1 1 0 1 1 0 0 0 1 0 0 0 0];

    [B1, A1] = get_SR(b1, 1);
    [B3, A3] = get_SR(b3, 3);
    [B5, A5] = get_SR(b5, 5);
    [B7, A7] = get_SR(b7, 7);
    [B9, A9] = get_SR(b9, 9);

end

function [B, A] = get_SR(b, n)
    g = [1 1 0 0 0 0 0 0 0 0 0 0 0 0 0];
    dim = 8;

    % Syndrome Machine 1: Division Remainder
    k = zeros(1, 15 + dim);
    B = [zeros(dim, 15)', eye(15)];
    for j = 0 : dim - 1
        B = 1 .* xor([k; B(1 : end - 1, :)], b' * B(end, :));
        B(1, dim - j) = 1;
    end
```

```
% Syndrome Machine 1: Remainder Evaluation
k = zeros(1, 15);
A = eye(15);
if n ~= 1
    for i = 2 : 15
        for j= 1 : (i - 1) * (n - 1)
            A(:, i) = 1 .* xor([0; A(1 : end - 1, i)],...
                          g' * A(end, i));
        end
    end
end
end
```

For the second method we need the matrices to get the syndromes.

INPUTS

- *dim*: the number of data bits encoded at the same time.
- *g*: the coefficients of the field generator polynomial.

OUTPUTS

- *S1, S3, S5, S7, S9*: the five matrices of dimension 15×23 used in the syndromes computation.

ROUTINE

```
function [S1, S3, S5, S7, S9] = get_SYN_HOR2()

    % Syndrome Machine 2: R(x) Evaluation with Horner
    S1 = get_SH(1);
    S3 = get_SH(3);
    S5 = get_SH(5);
    S7 = get_SH(7);
    S9 = get_SH(9);
end

function S = get_SH(n)
    g = [1 1 0 0 0 0 0 0 0 0 0 0 0 0 0 0];
    dim = 8;

    k = zeros(1, 15 + dim);
    S = [zeros(dim, 15)', eye(15)];
    for j = 1 : dim
        for i = 1 : n
            S = 1 .* xor([k; S(1 : end - 1, :)], g' * S(end, :));
        end
        S(1, dim - j + 1) = 1;
    end
end
```

D.3 Routine to get the matrix for the multiplier

The $GF(2^{15})$ multiplier used in the Berlekamp algorithm needs a matrix for its implementation, as described in Chap. 9 (Sect. 9.6).

INPUTS

- g: the coefficients of the field generator polynomial.

OUTPUTS

- P: the matrix of dimension 15×225 used to implement the multiplier.

ROUTINE

```
function [P] = get_PRODUCT()
    g = [1 1 0 0 0 0 0 0 0 0 0 0 0 0 0];
    k = [0 0 0 0 0 0 0 0 0 0 0 0 0 0 0];

% A1 is the matrix to multiply by alfa
    A1 = eye(15);
    A1 = 1.* xor([k; A1(1:end-1,:)], g' * A1(end,:));

% Generation of P matrix
    E = eye(15);
    P = E
    for i = 1 : 14
        E = E * A1;
        P = [P, E];
    end

end
```

D.4 Routine to calculate the coefficients of the error locator polynomial

The routine executes the search of the coefficients of the error locator polynomial. In order to optimize the routine for the five errors case, the Berlekamp-Massey tree specific for the five errors case is used, as described in Chap. 9 (Sect. 9.6): all the routine variables are the ones defined in the same chapter – Eqs. (9.47) and (9.48). The C-routine is ready to be translated into the microcontroller language, because all the constraints imposed by the microcontroller are respected. For example, only 8 supporting registers are used and the operations sequence is the same as the one implemented in the microcontroller.

Once more, observe that not all the branches presented in Chap. 9 are implemented. This is because through a simulation it is possible to see that these branches are never accomplished or at least with a probability of less then 10^{-6}.

This implementations regards the custom Berlekamp tree presented in Fig. 9.29. The Berlekamp-Massey tree requires the additional routines to perform the operations in the Galois field. In order to see an implementation of the Galois field operations refer to Appendix B.

INPUTS

- *S1,S3,S5,S7,S9*: the syndromes.
- *err*: the error correcting capability of the code.
- *DIM_CAMPO*: the dimension of the working field, e.g. if we are working on GF(2^{15}) *DIM_CAMPO* is 15.

OUTPUTS

- *Lam1, lam2, lam3, lam4, lam5:* the coefficients of the error locator polynomial.
- *deg*: the degree of the error locator polynomial.

ROUTINE

```
void albero(void) {
    int i;

    if(sindro0() == 0)
        Reg8[0] = 0;
    else {
        if(verifica(S1) != 0) {
            for(i = 0; i < DIM_CAMPO; i++)
                Reg1[i] = S1[i];
            moltiplicazione(Reg1, S1, Reg1);
            moltiplicazione(Reg1, S1, Reg3);
            somma(Reg3, S3, Reg2);
/* We have calculated A */
            if(verifica(Reg2) == 0) {
                moltiplicazione(Reg1, Reg3, Reg4);
                somma(Reg4, S5, Reg3);
            /* We have calculated B */
                if(verifica(Reg3) == 0) {
                    moltiplicazione(Reg4, Reg1, Reg4);
                    somma(Reg4, S7, Reg5);
/* We have calculated N */
                    if(verifica(Reg5) == 0) {
                        moltiplicazione(Reg4, Reg1, Reg4);
                        somma(Reg4, S9, Reg3);
/* We have calculated X */
                        if(verifica(Reg3) == 0) {
                            printf("Only one error occurred\n");
                            for(i = 0; i < DIM_CAMPO; i++) {
                                lam1[i] = S1[i];
                                lam2[i] = 0;
                                lam3[i] = 0;
                                lam4[i] = 0;
                                lam5[i] = 0;
                                Reg8[0] = 1;
                            }
                        }
                    }
```

```
                            else {
                                printf("Uncorrectable error\n");
                                Reg8[0] = -1;
                            }
                        }
                        else {
                            printf("Uncorrectable error\n");
                            Reg8[0] = -1;
                        }
                    }
                    else  {
                        moltiplicazione(Reg3, Reg3, Reg5);
                        for(i = 0; i < DIM_CAMPO - 1; i++) {
                            moltiplicazione(Reg5, S1, Reg6);
                            moltiplicazione(Reg6, Reg6, Reg5);
                        }
                        moltiplicazione(Reg5, S3, Reg6);
                        for(i = 0; i < DIM_CAMPO; i++)
                            Reg2[i] = S3[i];
                        moltiplicazione(Reg2, Reg2, Reg2);
                        moltiplicazione(Reg2, S1, Reg2);
                        somma(Reg2, S7, Reg2);
                        somma(Reg6, Reg2, Reg6);
/* We have calculated O */
                        if(verifica(Reg6)!= 0)  {
                            moltiplicazione(Reg4, Reg1, Reg4);
                            moltiplicazione(Reg1, Reg4, Reg4);
                            somma(Reg4, S9, Reg4);
                            moltiplicazione(Reg6, Reg6, Reg7);
                            for(i = 0; i < DIM_CAMPO - 1; i++)  {
                                moltiplicazione(Reg3, Reg7, Reg8);
                                moltiplicazione(Reg8, Reg8, Reg7);
                            }
                            moltiplicazione(Reg2, Reg7, Reg2);
                            somma(Reg4, Reg2, Reg4);
                            moltiplicazione(Reg5, S5, Reg2);
                            somma(Reg2, Reg4, Reg4);
/* We have calculated Q */
                            if(verifica(Reg4) == 0) {
                                printf("4 errors occurred\n");
                                for(i = 0; i < DIM_CAMPO; i++) {
                                    lam1[i] = S1[i];
                                    lam2[i] = Reg7[i];
                                    lam4[i] = Reg5[i];
                                    lam5[i] = 0;
                                }
                                moltiplicazione(Reg7, S1, lam3);
                                Reg8[0] = 4;
                            }
                            else  {
                                printf("5 errors occurred\n");
                                moltiplicazione(Reg4, Reg4, Reg8);
                                for(i = 0; i < DIM_CAMPO - 1; i++)  {
                                    moltiplicazione(Reg8, Reg3, Reg6);
                                    moltiplicazione(Reg6, Reg6, Reg8);
                                }
                                for(i = 0; i < DIM_CAMPO; i++)  {
                                    lam1[i] = S1[i];
```

```
                                    lam2[i] = Reg7[i];
                                }
                                moltiplicazione(Reg7, S1, lam3);
                                somma(Reg5, Reg8, lam4);
                                moltiplicazione(Reg8, S1, lam5);
                                Reg8[0] = 5;
                            }
                        }
                        else  {
                            printf("Uncorrectable error\n");
                            Reg8[0] = -1;
                        }
                    }
                }
                else  {
                    moltiplicazione(Reg2, Reg2, Reg4);
                    for(i = 0; i < DIM_CAMPO - 1; i++)  {
                        moltiplicazione(Reg4, S1, Reg5);
                        moltiplicazione(Reg5, Reg5, Reg4);
                    }
                    moltiplicazione(Reg3, Reg1, Reg5);
                    somma(Reg5, S5, Reg6);
                    moltiplicazione(Reg4, S3, Reg7);
                    somma(Reg6, Reg7, Reg6);
/* We have calculated C */
                    if(verifica(Reg6)==0)  {
                        for(i = 0; i < DIM_CAMPO; i++)
                            Reg1[i] = S3[i];
                        moltiplicazione(Reg1, S3, Reg1);
                        moltiplicazione(Reg1, S1, Reg1);
                        somma(Reg1, S7, Reg1);
                        moltiplicazione(Reg4, S5, Reg6);
                        somma(Reg6, Reg1, Reg1);
/* We have calculated R */
                        if(verifica(Reg1) == 0)  {
                            moltiplicazione(Reg4, S7, Reg6);
                            moltiplicazione(Reg5, Reg3, Reg5);
                            moltiplicazione(Reg5, S1, Reg5);
                            somma(Reg5, S9, Reg5);
                            somma(Reg5, Reg6, Reg5);
/* We have calculated S */
                            if(verifica(Reg5) == 0)  {
                                printf("2 errors occurred");
                                for(i = 0; i < DIM_CAMPO; i++)  {
                                    lam1[i] = S1[i];
                                    lam2[i] = Reg4[i];
                                    lam3[i] = 0;
                                    lam4[i] = 0;
                                    lam5[i] = 0;
                                }
                                Reg8[0] = 2;
                            }
                            else  {
                                printf("Uncorrectable error\n");
                                Reg8[0] = -1;
                            }
                        }
                        else  {
```

```
                        moltiplicazione(Reg5, Reg3, Reg3);
                        moltiplicazione(Reg3, S1, Reg3);
                        somma(Reg3, S9, Reg3);
                        moltiplicazione(Reg4, S7, Reg6);
                        somma(Reg3, Reg6, Reg3);
                        moltiplicazione(Reg1, Reg1, Reg6);
                        for(i = 0; i < DIM_CAMPO - 1; i++)  {
                            moltiplicazione(Reg6, Reg2, Reg7);
                            moltiplicazione(Reg7, Reg7, Reg6);
                        }
                        somma(Reg5, S5, Reg5);
                        moltiplicazione(Reg5, Reg6, Reg5);
                        somma(Reg3, Reg5, Reg3);
/* We have calculated T */
                        if(verifica(Reg3)!= 0)  {
                            printf("5 errors occurred\n");
                            moltiplicazione(Reg3, Reg3, Reg7);
                            for(i = 0; i < DIM_CAMPO - 1; i++)  {
                                moltiplicazione(Reg7, Reg1, Reg8);
                                moltiplicazione(Reg8, Reg8, Reg7);
                            }
                            somma(Reg4, Reg7, lam2);
                            moltiplicazione(Reg7, S1, lam3);
                            moltiplicazione(Reg7, Reg4, Reg8);
                            somma(Reg8, Reg6, lam4);
                            moltiplicazione(Reg6, S1, lam5);
                            for(i = 0; i < DIM_CAMPO; i++)
                                lam1[i] = S1[i];
                            Reg8[0] = 5;
                        }
                        else  {
                            printf("Uncorrectable error\n");
                            Reg8[0] = -1;
                        }
                    }
                }
                else  {
                    moltiplicazione(Reg6, Reg6, Reg7);
                    for(i = 0; i < DIM_CAMPO - 1; i++)  {
                        moltiplicazione(Reg7, Reg2, Reg8);
                        moltiplicazione(Reg8, Reg8, Reg7);
                    }
                    for(i = 0; i < DIM_CAMPO; i++)
                        Reg1[i] = S3[i];
                    moltiplicazione(Reg1, S3, Reg1);
                    moltiplicazione(Reg1, S1, Reg2);
                    somma(Reg2, S7, Reg1);
                    moltiplicazione(Reg7, Reg5, Reg8);
                    somma(Reg1, Reg8, Reg1);
                    somma(Reg7, Reg4, Reg8);
                    moltiplicazione(Reg8, S5, Reg8);
                    somma(Reg8, Reg1, Reg1);
/* We have calculated U */
                        if(verifica(Reg1) == 0)  {
                            moltiplicazione(Reg5, Reg3, Reg1);
                            moltiplicazione(Reg1, S1, Reg1);
                            somma(Reg1, S9, Reg1);
                            moltiplicazione(Reg7, Reg2, Reg8);
```

```
                        somma(Reg8, Reg1, Reg8);
                        somma(Reg4, Reg7, Reg3);
                        moltiplicazione(Reg3, S7, Reg4);
                        somma(Reg4, Reg8, Reg8);
/* We have calculated V */
                        if(verifica(Reg8) == 0)  {
                            printf("3 errors occurred\n");
                            moltiplicazione(Reg7, S1, lam3);
                            for(i = 0; i < DIM_CAMPO; i++) {
                                lam4[i] = 0;
                                lam5[i] = 0;
                                lam2[i] = Reg3[i];
                                lam1[i] = S1[i];
                            }
                            Reg8[0] = 3;
                        }
                        else {
                            printf("Uncorrectable error\n");
                            Reg8[0] = -1;
                        }
                    }
                    else  {
                        moltiplicazione(Reg5, Reg3, Reg5);
                        moltiplicazione(Reg5, S1, Reg5);
                        somma(Reg5, S9, Reg5);
                        moltiplicazione(Reg1, Reg1, Reg8);
                        for(i = 0; i < DIM_CAMPO - 1; i++)  {
                            moltiplicazione(Reg6, Reg8, Reg3);
                            moltiplicazione(Reg3, Reg3, Reg8);
                        }
                        somma(Reg7, Reg8, Reg6);
                        moltiplicazione(Reg2, Reg6, Reg2);
                        somma(Reg5, Reg2, Reg5);
                        somma(Reg6, Reg4, Reg3);
                        moltiplicazione(Reg3, S7, Reg2);
                        somma(Reg5, Reg2, Reg5);
                        moltiplicazione(Reg8, Reg4, Reg2);
                        moltiplicazione(Reg2, S5, Reg4);
                        somma(Reg5, Reg4, Reg5);
/* We have calculated W */
                        if(verifica(Reg5) ==0 ) {
                            printf("4 errors occurred\n");
                            moltiplicazione(Reg6, S1, lam3);
                            for(i = 0; i < DIM_CAMPO; i++)  {
                                    lam1[i] = S1[i];
                                    lam2[i] = Reg3[i];
                                    lam4[i] = Reg2[i];
                                    lam5[i] = 0;
                            }
                            Reg8[0] = 4;
                        }
                        else {
                            printf("5 errors occurred\n");
                            moltiplicazione(Reg5, Reg5, Reg4);
                            for(i = 0; i < DIM_CAMPO - 1; i++)  {
                                moltiplicazione(Reg4, Reg1, Reg5);
                                moltiplicazione(Reg5, Reg5, Reg4);
                            }
```

```
                              for(i = 0; i < DIM_CAMPO; i++)
                                  lam1[i] = S1[i];
                              somma(Reg3, Reg4, lam2);
                              somma(Reg4, Reg6, Reg5);
                              moltiplicazione(Reg5, S1, lam3);
                              somma(Reg6, Reg3, Reg3);
                              somma(Reg3, Reg7, Reg6);
                              moltiplicazione(Reg6, Reg4, Reg6);
                              moltiplicazione(Reg3, Reg8, Reg8);
                              somma(Reg6, Reg8, lam4);
                              moltiplicazione(Reg7, Reg4, Reg5);
                              moltiplicazione(Reg5, S1, lam5);
                              Reg8[0] = 5;
                          }
                      }
                  }
              }
          }
          else  {
              if(verifica(S3) != 0) {
                  if(verifica(S5) != 0) {
                      for(i = 0; i < DIM_CAMPO; i++)
                          Reg1[i] = S5[i];
                      moltiplicazione(Reg1, Reg1, Reg2);
                      for(i = 0; i < DIM_CAMPO - 1; i++)   {
                          moltiplicazione(Reg2, S3, Reg1);
                          moltiplicazione(Reg1, Reg1, Reg2);
                      }
                      moltiplicazione(Reg2, S5, Reg1);
                      somma(Reg1, S7, Reg1);
/* We have calculated J */
                      if(verifica(Reg1) == 0) {
                          moltiplicazione(Reg2, S7, Reg3);
                          for(i = 0; i < DIM_CAMPO; i++)
                              Reg4[i] = S3[i];
                          moltiplicazione(Reg4, Reg4, Reg4);
                          moltiplicazione(Reg4, S3, Reg4);
                          somma(Reg3, Reg4, Reg3);
                          somma(Reg3, S9, Reg3);
/* We have calculated K */
                          if(verifica(Reg3) == 0) {
                              printf("3 errors occurred\n");
                              for(i = 0; i < DIM_CAMPO; i++) {
                                  lam1[i] = 0;
                                  lam2[i] = Reg2[i];
                                  lam3[i] = S3[i];
                                  lam4[i] = 0;
                                  lam5[i] = 0;
                              }
                              Reg8[0] = 3;
                          }
                          else   {
                              printf("Uncorrectable error\n");
                              Reg8[0] = -1;
                          }
                      }
                      else   {
                          moltiplicazione(Reg1, Reg1, Reg3);
```

```
                        for(i = 0; i < DIM_CAMPO - 1; i++)  {
                            moltiplicazione(Reg3, S3, Reg4);
                            moltiplicazione(Reg4, Reg4, Reg3);
                        }
                        moltiplicazione(Reg2, S7, Reg4);
                        somma(Reg4, S9, Reg4);
                        moltiplicazione(Reg3, S5, Reg5);
                        somma(Reg4, Reg5, Reg4);
                        for(i = 0; i < DIM_CAMPO; i++)
                            Reg6[i] = S3[i];
                        moltiplicazione(Reg6, Reg6, Reg6);
                        moltiplicazione(Reg6, S3, Reg6);
                        somma(Reg4, Reg6, Reg4);
/* We have calculated L */
                        if(verifica(Reg4) == 0) {
                            printf("4 errors occurred\n");
                            for(i = 0; i < DIM_CAMPO; i++) {
                                lam1[i] = 0;
                                lam2[i] = Reg2[i];
                                lam3[i] = S3[i];
                                lam4[i] = Reg3[i];
                                lam5[i] = 0;
                            }
                            Reg8[0] = 4;
                        }
                        else  {
                            printf("5 errors occurred\n");
                            for(i = 0; i < DIM_CAMPO; i++) {
                                lam1[i] = 0;
                                lam3[i] = S3[i];
                            }
                            moltiplicazione(Reg4, Reg4, Reg5);
                            for(i = 0; i < DIM_CAMPO - 1; i++) {
                                moltiplicazione(Reg5, Reg1, Reg4);
                                moltiplicazione(Reg4, Reg4, Reg5);
                            }
                            moltiplicazione(Reg5, S3, lam5);
                            somma(Reg5, Reg2, lam2);
                            moltiplicazione(Reg2, Reg5, Reg4);
                            somma(Reg4, Reg3, lam4);
                            Reg8[0] = 5;
                        }
                    }
                }
                else {
                    printf("Uncorrectable error\n");
                    Reg8[0] = -1;
                }
            }
            else {
                printf("Uncorrectable error\n");
                Reg8[0] = -1;
            }
        }
    }
    deg = Reg8[0];
}
```

D.5 Routine to get matrices for the Chien machine

The Chien machine needs the matrices for the evaluation of the error locator polynomial in the elements of the field. Since we are considering codes corrector of five errors we need five matrices, one for each coefficient of the error locator polynomial.

INPUTS

- *g*: the coefficients of the field generator polynomial.

OUTPUTS

- *A1, A2, A3, A4, A5*: the matrices of dimension 15 × 15 used in the Chien machines described in Chap. 9.

ROUTINE

```
function [A1, A2, A3, A4, A5] = get_CHIEN()

    A1 = get_CH(1);
    A2 = get_CH(2);
    A3 = get_CH(3);
    A4 = get_CH(4);
    A5 = get_CH(5);
end

function A = get_CH(n)

    g = [1 1 0 0 0 0 0 0 0 0 0 0 0 0 0];
    k = [0 0 0 0 0 0 0 0 0 0 0 0 0 0 0];
    A = eye(15);
    for i = 1 : n
        A = 1 .* xor([k; A(1 : end - 1, :)], g' * A(end, :));
    end
end
```

D.6 Global matrix optimization for the Chien machine

Another approach is the global matrix optimization. Suppose we want to analyze 4 bits in parallel at every step. We will group in a single matrix each matrix with the same inputs, in order to have five matrices, one for each *Lambda*'s coefficients. The first step is analyzing these five matrices to find common terms among columns. These common terms represent new variables to be used in computation.

The following Matlab routine finds common terms among columns and then finds a new matrix optimized with the new variables introduced.

As it is possible to see in the following, a lot of subroutines are introduced. For every subroutine a comment at the beginning indicates its functionality.

At every step, the Chien machine will have to multiply a vector containing the registers content and the new data with a matrix, in order to obtain new variables. Then, this vector of new variables will be multiplied by a an optimized matrix to obtain the evaluated polynomial at that step of the decoding algorithm.

This kind of optimization is exactly the same as it would be done by synthesis tools.

INPUTS

- *dim_campo*: the dimension of the working field, e.g. if we are working on $GF(2^{15})$ *dim_campo* is 15.
- *alpha1, alpha2, alpha3, alpha4, alpha6, alpha8, alpha9, alpha12, alpha16, alpha5, alpha10, alpha15, alpha20, alpha7, alpha11, alpha13, alpha14:* all the matrices needed to perform multiplication with the chosen parallelism.

OUTPUTS

- *nuovelam1, nuovelam2, nuovelam3, nuovelam4, nuovelam5*: the five matrices introducing the new variables.
- *ottimolam1, ottimolam2, ottimolam3, ottimolam4, ottimolam5*: the five matrices optimized.

ROUTINE

```
function [nuovelam1, ottimolam1, nuovelam2, ottimolam2, ...
          nuovelam3, ottimolam3, nuovelam4, ottimolam4, ...
          nuovelam5, ottimolam5,] = ottimizzo(dim_campo)

    global dim_campo
    dim_campo = 15;

    load alpha1.txt -ascii
    load alpha2.txt -ascii
    load alpha3.txt -ascii
    load alpha4.txt -ascii
    load alpha6.txt -ascii
    load alpha8.txt -ascii
    load alpha9.txt -ascii
    load alpha12.txt -ascii
    load alpha16.txt -ascii
    load alpha5.txt -ascii
    load alpha10.txt -ascii
    load alpha15.txt -ascii
    load alpha20.txt -ascii


    A = [alpha1; alpha2; alpha3; alpha4];
    I = calcolo_indici(A);
    numero = conto_coppie(I);
    estraggo = estrazione(numero);
    combino = combinazioni(estraggo);
    nuovelam1 = nuove_variabili(combino);
```

```
    ottimolam1 = matrice_ottima(combino,A);

    A = [alpha2; alpha4; alpha6; alpha8];
    I = calcolo_indici(A);
    numero = conto_coppie(I);
    estraggo = estrazione(numero);
    combino = combinazioni(estraggo);
    nuovelam2 = nuove_variabili(combino);
    ottimolam2 = matrice_ottima(combino, A);

    A = [alpha3; alpha6; alpha9; alpha12];
    I = calcolo_indici(A);
    numero = conto_coppie(I);
    estraggo = estrazione(numero);
    combino = combinazioni(estraggo);
    nuovelam3 = nuove_variabili(combino);
    ottimolam3 = matrice_ottima(combino, A);

    A = [alpha4; alpha8; alpha12; alpha16];
    I = calcolo_indici(A);
    numero = conto_coppie(I);
    estraggo = estrazione(numero);
    combino = combinazioni(estraggo);
    nuovelam4 = nuove_variabili(combino);
    ottimolam4 = matrice_ottima(combino, A);

    A = [alpha5; alpha10; alpha15; alpha20];
    I = calcolo_indici(A);
    numero = conto_coppie(I);
    estraggo = estrazione(numero);
    combino = combinazioni(estraggo);
    nuovelam5 = nuove_variabili(combino);
    ottimolam5 = matrice_ottima(combino, A);

return;

%
% The routine finds the position of the ones for each column
%
function I = calcolo_indici(matrix)
    global dim_campo
    somma = sum(matrix');
    somma2 = sum(somma > 1);
    I = zeros(dim_campo, 3);
    for i = 1 : 4 * dim_campo
        if(somma(i) > 1)
            for k = 1 : somma(i)
                for j = 1 : dim_campo
                    if(matrix(i, j) == 1)
                        if(k == 1)
                            indice(k) = j;
                            I(i, k) = j;
                            break;
                        end;
                        if(k > 1)
                            if(k > 2)
                                if((j ~= indice(k - 1)) & ...
                                    (j ~= indice(k - 2)))
```

```
                                        indice(k) = j;
                                        I(i, k) = j;
                                    end;
                                else
                                    if (j ~= indice(k - 1))
                                        indice(k) = j;
                                        I(i, k) = j;
                                        break;
                                    end;
                                end;
                            end;
                        end;
                    end;
                end;
            end;
        end;
    end;
end;


%
% The routine finds how many times a given pair of 1 is repeated
% in the matrix
%
function aaa = conto_coppie(M)
    global dim_campo
    numero = zeros(105, 8);
    riga = 1;
    for n = 1 : dim_campo
        for m = n + 1 : dim_campo
            indice1 = n;
            indice2 = m;
            for i = 1 : size(M, 1)
                for k = 1 : 3
                    numero(riga, 1) = indice1;
                    numero(riga, 2) = indice2;
                    if(M(i, 1) == 0)
                        break;
                    end;
                    for h = 1 : 3
                        if((M(i, k) == indice1) & ...
                            (M(i, h) == indice2))
                            numero(riga, 3) = numero(riga, 3) + 1;
                            numero(riga, 3 + numero(riga, 3)) = i;
                        end;
                    end;
                end;
            end;
            riga = riga + 1;
        end;
    end;
end;


%
% For each possible couple of ones, the routine finds those ones
% in the matrix and indicates their column position
%
function estraggo=estrazione(mat)
    rigaok = 1;
    for i = 1 : 105
```

```
        if (mat(i, 3) > 1)
            estraggo(rigaok, :) = mat(i, :);
            rigaok = rigaok + 1;
        end;
    end;
end;


%
% The routine finds new variables. There is the constraint of only
% one new variable per column.
%
function combino = combinazioni(matr)
    rig = 1;
    for i = 1 : size(matr, 1)
        flag(i) = 0;
        for k = i - 1 : -1 : 1
            for j = 4 : 8
                for h = 4 : 8
                    if((matr(i, j) ~= 0) & ...
                      (matr(i, j) == matr(k, h)))
                        flag(i) = 1;
                        if(matr(i, 3) > matr(k, 3))
                            flag(i) = 0;
                            flag(k) = 1;
                        end;
                    end;
                end;
            end;
        end;
    end;
    for i = 1 : size(matr, 1)
        if (flag(i) == 0)
            combino(rig, :) = matr(i, :);
            rig = rig + 1;
        end;
    end;
end;


function nuove = nuove_variabili(comb)
    global dim_campo
    nuove = zeros(dim_campo, size(comb, 1));
    for i = 1 : size(comb, 1)
        nuove(comb(i, 1), i) = 1;
        nuove(comb(i, 2), i) = 1;
    end;
end;


%
% The routine optimizes a matrix with the new variables introduced.
%
function ottimo = matrice_ottima(comb, matrix)
    global dim_campo
    ottimo = zeros(4 * dim_campo, dim_campo + size(comb, 1));
    ottimo(:, 1 : dim_campo) = matrix;
    for i = 1 : size(comb, 1)
        for j = 4 : 8
            if (comb(i, j) ~= 0)
                ottimo(comb(i, j), comb(i, 1)) = 0;
```

```
                    ottimo(comb(i, j), comb(i, 2)) = 0;
                    ottimo(comb(i, j), dim_campo + i) = 1;
                end;
            end;
        end;
    end;
```

D.7 BCH flow overview

The following Matlab program uses the previously calculated matrices to develop the BCH encoding and decoding, as emulation of the hardware implementation.

The *BCH_flow* describes the main flow. The message *m* is a random message of 2102 Bytes. The parity *p* is calculated with the *BCH_PAR* function and added to the message *m* to give the codeword *c*. Some errors could be summed to this codeword. The resultant received message *r* follows the decoding path. First of all, the function *BCH_SYN* is used to compute syndromes. If all the syndromes are not equal to zero, the Berlekamp function and the Chien function will be used.

ROUTINE

```
function BCH_flow()

    len_mess = 2102;
    dim_word = 8;

    % Generate random data vector
    m = (rand(len_mess, dim_word) > .5) * 1.0;

    % Parity computation
    p = BCH_PAR(m);

    % c: Stored code word
    c = [m; p];

    % Force errors
    e = zeros(size(c));
    e(1, 1) = 1;
    e(3, 1) = 1;
    e(5, 1) = 1;
    e(7, 1) = 1;

    % r: Read message
    r = xor(c, e);

    % Syndromes calculation
    [S1, S2, S3, S4, S5] = BCH_SIN(r)

    num_err = 0;
    if sum([S1, S2, S3, S4, S5]) ~= 0
        [num_err, L1, L2, L3, L4, L5] = Berlekamp(S1, S2, S3, S4, S5)
        [Err_Pos, Err_Mask] = Chien(num_err, len_mess + 10, L1,L2,...
                              L3, L4, L5)
    end
end
```

The *BCH_PAR* function gets the 75-bit parity vector and generates the parity matrix 10×8 filling the empty bits of the last byte with zeros.

INPUTS

- *m*: array 2102×8 with the message to be encoded.

OUTPUTS

- *c*: the matrix 10×8 with the parity of the message.

ROUTINE

```
function c = BCH_PAR(m)
    PAR = get_PAR();

    c = zeros(1,75);
    for i = (1:size(m,1))
        c = rem((PAR * [fliplr(m(i,:)) c]')',2);
    end
    c = reshape(fliplr([0, 0, 0, 0, 0, c]), [8 10])'
end
```

The *BCH_SYN* function gets syndromes from the received message. As for the hardware implementation, the received message is aligned with the end of the word.

INPUTS

- *r*: array 2112×8 with the received message.

OUTPUTS

- *S1, S3, S5, S7, S9*: the calculated syndromes.

ROUTINE

```
function [S1, S3, S5, S7, S9] = BCH_SYN(r)

    [M1, M3, M5, M7, M9] = get_SYN_HOR();

    tmp = r';
    tmp = [0, 0, 0, 0, 0, tmp(1 : end - 5)];
    r = reshape(tmp, size(r'))';

    S1 = zeros(1, 15);
    S3 = zeros(1, 15);
    S5 = zeros(1, 15);
    S7 = zeros(1, 15);
    S9 = zeros(1, 15);
```

```
    for i = (1 : size(r, 1))
        S1 = rem((M1 * [fliplr(r(i, :)) S1]')', 2);
        S3 = rem((M3 * [fliplr(r(i, :)) S3]')', 2);
        S5 = rem((M5 * [fliplr(r(i, :)) S5]')', 2);
        S7 = rem((M7 * [fliplr(r(i, :)) S7]')', 2);
        S9 = rem((M9 * [fliplr(r(i, :)) S9]')', 2);
    end
end
```

Appendix E: Erasure decoding technique

In this Appendix a possible C-language implementation of the *BCHerasure* routine described in Chap. 10 is presented.

The routine is called after a decoding failure of the classical BCH decoding method. It analyzes the input message building another one with erasures to be decoded. After substituting zeroes in the erasure positions, it makes use of the BCH decoding methods described in the previous appendices.

The routine needs some subroutines. Also these supporting subroutines are described here. For each routine there is a brief description of its functionality, its inputs and its outputs.

E.1 Subroutines

The first routine analyzes every single bit, in order to understand if the read bit is a 0, a 1 or an erasure.

INPUTS

- Nl: the number of readings.
- Ns: the threshold used in the decision. Observe that we use only one threshold because we consider a symmetrical error probability.
- s: the sum of read data for every cell.

OUTPUTS

- r: the data value for the bit.
1. $r=0$ if $s \leq Ns$;
2. $r=1$ if $s \geq Nl-Ns$;
3. $r=2$ otherwise. This is the erasure case.

ROUTINE

```
int recdata(int Nl, int Ns, int s) {
    int r;
```

```
    r = 0;
    if(s <= Ns)
        r = 0;
    else if (s >= Nl - Ns)
        r = 1;
    else
        r = 2;
    return r;
}
```

The following routine builds the message containing the erasures to be decoded.

INPUTS

- *sm*: the pointer to the vector containing the sum of the read messages.
- *length:* the codeword length.
- *Nl*: the number of readings.
- *Ns:* the threshold used in the decision. Observe that we use only one threshold because we consider a symmetrical error probability.

OUTPUTS

- *sm*: the pointer to the vector containing the message to be decoded.

ROUTINE

```
void rec(int *sm, int length, int Nl, int Ns) {
    int i;

    for (i = 0; i < length; i++)
        sm[i] = recdata(Nl, Ns, sm[i]);
    return;
}
```

The next routine substitutes zeroes or ones in the erasure positions.

INPUTS

- *length:* the codeword length.
- *x*: message to be decoded.
- *e*: the substituting value, that is a 0 or a 1.

OUTPUTS

- *xe*: the message *x* with the erasures substituted with *e*.

ROUTINE

```
void substitution0(int *sm, int length, int e, int x[]) {
    int i;

    for(i = 0; i < length; i++) {
        x[i] = sm[i];
        if (x[i] == 2)
            x[i] = e;
    }
    return;
}
```

Now a routine that decodes the message with the erasures (substituted with zeroes or ones) is described. The routine calls the subroutine *decode* representing the classical BCH decoding method.

INPUTS

- *length:* the codeword length.
- **sm*: the pointer to the vector containing the message to be decoded.
- *e*: the substituting value, that is a 0 or a 1.

OUTPUTS

- *successo0*: it tells if the decoding procedure was successful.

ROUTINE

```
int decode0(int *sm, int length, int e) {
    int successo0, received0[length], i;

    successo0=-1;
    for(i = 0; i < length; i++)
        received0[i]=0;
    substitution0(sm, length, e, received0);
    successo0 = decode(received0, length);
    /* successo0=0 if decoding is successful otherwise it is 1 */
    if (successo0 == 0) {
        for(i = 0; i < length; i++)
            sm[i] = received0[i];
    }
    return successo0;
}
```

Finally, the routine implementing the binary erasure decoding technique follows.

INPUTS

- *length:* the codeword length.
- **sm*: the pointer to the vector containing the message to be decoded.

OUTPUTS

- *sm*: the decoded message. If the decoding procedure fails it is composed by all values equal to "-1".

ROUTINE

```
int erasure(int *sm,int length) {
    int successo0, i;

    successo0 = -1;
    successo0 = decode0(sm, length, 0);
    if(successo0 == 1) successo0 = decode0(sm, length, 1);
    if(successo0 == 1)
        for(i = 0; i < length; i++)
            sm[i] = -1;
    return successo0;
}
```

E.2 Erasure decoding routine

At the end the main routine of the decoding.

INPUTS

- *dim:* the data length in bytes.
- *PARITA*: the number of parity bits.
- **sm*: the pointer to the vector containing the sum of the read messages.
- *Nl*: the number of readings.
- *Ns:* the threshold used in the decision. Observe that we use only one threshold because we consider a symmetrical error probability.

OUTPUTS

- *sm*: the decoded message. If the decoding procedure fails it is composed by all values equal to "-1".

ROUTINE

```
void BCHerasure(int *sm, int dim, int N1, int Ns) {
    int successo, i, length;

    successo = -1;
    length = (dim * 8) + PARITA;
    rec(sm, length, N1, Ns);
    successo = erasure(sm, length);
    if(successo != 0)
        printf("Decoding failure\n");
    return;
}
```

Subject index

.

CPSIA information can be obtained at www.ICGtesting.com
Printed in the USA
LVOW05*0904161114

413959LV00002B/154/P